Total Eclipses:
Science, Observations, Myths and Legends

Springer
London
Berlin
Heidelberg
New York
Barcelona
Hong Kong
Milan
Paris
Santa Clara
Singapore
Tokyo

Pierre Guillermier and Serge Koutchmy

Total Eclipses: Science, Observations, Myths and Legends

Dr Pierre Guillermier
Member of the Société Astronomique de France
3 rue Beethoven, 75016 Paris, France

Dr Serge Koutchmy
Directeur de Recherche
Institut d'Astrophysique de Paris (CNRS)
Paris, France

Original French edition, *Eclipses Totales: Histoire, Découvertes, Observations*
Published by © MASSON editeur, Paris, 1998

This work has been published with the help of the French Ministère de la Culture—Centre national du livre

Translator: Bob Mizon, 38 The Vineries, Colehill, Wimborne, Dorset, BH21 2PX

SPRINGER–PRAXIS SERIES IN ASTRONOMY AND ASTROPHYSICS
SERIES EDITOR: John Mason B.Sc., Ph.D.

ISBN 1-85233-160-7 Springer-Verlag Berlin Heidelberg New York

British Library Cataloguing in Publication Data

Library of Congress Cataloging-in-Publication Data
Guillermier, Pierre, 1960–
Total eclipses : science, observations, myths, and legends /
Pierre Guillermier.
p. cm. – (Springer–Praxis series in astronomy and astrophysics)
Includes bibliographical references and index.
ISBN 1-85233-169-7 (alk. paper)
1. Solar eclipses. I. Title. II. Series.
QB541.G85 1999
523.7′8–dc21 99-19976
CIP

Printed by MPG Books Ltd, Bodmin, Cornwall, UK

Cover design: Jim Wilkie
Typesetting: BookEns Ltd, Royston, Herts. UK

Printed on acid-free paper supplied by Precision Publishing Papers Ltd, UK

Table of contents

Foreword to the English edition

The last total solar eclipse of the millennium, on 11 August 1999, has aroused immense interest in the subject of solar eclipses in general. As far as mainland Britain is concerned, this eclipse marks the end of a 72-year interval since the previous total eclipse, and there will not be another until September 2090. Prompted by this heightened awareness of the eclipse phenomenon, many new books on the subject have been published, but almost all of these are either very elementary or else highly technical. This new book, the result of amateur and professional collaboration, provides a most useful link between the two types. Neither is it confined to the August 1999 eclipse; there are many fascinating historical notes, for example, and there is also a mathematical appendix, together with an extensive reference list for those who wish to take their studies further. Certainly I feel that this book makes an excellent and valuable contribution to the literature on solar eclipses.

Patrick Moore
Selsey, England, May 1999

Preface

Marvellous, fantastic, extraordinary, unforgettable, magical ... we soon exhaust the adjectives when describing a total eclipse of the Sun.

Truly, such a spectacle is awe-inspiring, although eclipses are frequent events. The Moon's shadow may sweep across the Earth's surface several times a year, but for a given place, only two or three times a century – or even fewer, the closer one lives to the poles. So the wondrous sight of a total eclipse is indeed a rare thing.

For my part, I have seen only one eclipse, from Khartoum in the Sudan in 1952. Its beauty was such that, for fear of anticlimax, I did not feel minded to see another, even that of 1961 at Haute-Provence Observatory. In Khartoum, as totality arrived, Mercury, Venus and Mars were visible, along with a few stars, and I could make out the Milky Way. Especially wonderful was the 'solar minimum' corona, with three long, narrow streamers, visible to the naked eye out to a considerable distance from the disk. The weather was perfect, and the sky very clear. Far off, near the horizon, the sky remained bright in all directions. Total silence reigned.

In all conditions, from deserts and mountains, from aircraft and ships, Serge Koutchmy, with his usual energy, and Pierre Guillermier, a passionate amateur, have in their turn witnessed many total eclipses in the company of their colleagues. In this book they present to us a plethora of superb images of the solar corona drawn from their eclipse observations.

Here is a timely book, as the path of the eclipse of August 1999 crosses Europe. This is the first total eclipse to be seen from England since 1927, and from France since 1961, and is exceptional as the last of this millennium! The authors provide the observer of this superb event with an in-depth introduction to the physics of the Sun, and to the methodical observation of eclipses. We are expertly told how science derives information from such observations.

It must be remembered that, until Lyot invented the coronagraph in 1930, it was possible to observe the Sun's corona only during those fleeting minutes of total eclipse, and an astronomer could attend very few of them. Eclipses have been studied since ancient times, and more especially during the nineteenth and twentieth centuries on many a far-flung expedition with all its risks and dangers; but nowadays, these trips, with greater numbers of participants, are well coordinated, and reach places once thought inaccessible, where skies are pure and clear. Equipment is ever more elaborate and sophisticated, and with it, much time has been spent in trying to determine the origin of the corona. Could it be the lunar atmosphere? Was it cold or hot? Gaseous or dusty?

Why, one may ask, is it still worth mounting such costly eclipse expeditions, with their attendant complex organization and uncertain outcome, when the corona can nowadays be continuously observed by machines like the European SOHO satellite or Japan's Yohkoh, or with a coronagraph from observatories like the Pic-du-Midi? The simple answer is that, in the case of a coronagraph, with its occulting disc only a few centimetres wide, the surrounding sky will appear bright close in to the Sun, and only the inner corona will be visible.

Also, those excellent satellites SOHO and Yohkoh concentrate rather on X-rays, the infrared and the ultraviolet, and their working lives will depend upon costs. They are complementing, rather than competing with, ground-based observations of the Sun made over many decades, in an homogeneous way which allows study of the corona over several cycles of solar activity. We now see why energetic scientists like Serge Koutchmy, in conjunction with experienced amateurs such as Pierre Guillermier, have crossed the five continents, and even oceans, in order not to miss any total eclipse of the Sun. His experience endows this book with great authenticity and authority, stemming from his great competence as an observer and tireless traveller.

Even the non-scientific among us can learn a lot from the passage of the Moon's shadow across the Earth's surface. For example, it provides an insight into the accuracy of which astronomical predictions are capable: the eclipse track can be calculated to a fraction of a second or to the nearest metre – a striking illustration of the exact character of scientific deduction.

For all of us, whether we are professional or amateur solar astronomers, there remains the simple sentiment of the beauty of celestial phenomena, as seen from the comparatively dull pebble called Earth.

Jean-Claude Pecker
Honorary Professor at the Collège de France
Member of the French Academy of Sciences

Acknowledgements

We extend our warmest thanks to Jean-Claude Pecker for the preface; Jean Aboudarham, Karen Chalonge, Georges Ecochard, Roger Ferlet, Violaine Guillermier, Valéry Koutchmy, Hervé Roux-Mayoud, Jay Pasachoff, Denys Samain and Michel Sarrazin for all their help with the various stages of the manuscript; Gilles Haéri, Jean-Luc Blanc and their team for their support in the realisation of this project; and Bob Mizon (translator), Bob Marriott (editor) and Clive Horwood (Chairman of Praxis).

Pierre Guillermier would like to thank Yves Méziat-Burdin and Maurice Marrel, with whom he has shared many observations; Alice and Paul, who put up with 'his eclipses' as he travelled the world in search of the Moon's shadow.

Serge Koutchmy would have liked to thank by name all those who accompanied him on one or more eclipse trips, and who have done much to bring into existence his own observations; but these individuals are far too numerous, and he fears that his memory might let him down. He cannot, however, forget those who have, sadly, already left us: Roland Caron, Daniel Chalonge, Marius Laffineur, Alexei Nesmjanovich, Gordon Newkirk, Gennady Nikolsky, Sergei Vsekhsvjatsky – to all of whom he dedicates his contribution to this book.

Introduction

Every day, a familiar and dazzling yellow ball, the Sun, moves across the sky, providing us with heat and light, and indeed, our very existence. In summer, we complain if its rays are too hot. In winter, we pine for its presence. In spite of all our complaints, we are grateful for its constancy, and we have grown used to the rhythm of days, nights and seasons which it imposes upon us.

Outside this customary picture, what do we know about our star? How does it work? What will happen to it in the future? What secrets does it still keep from the astrophysicists?

As the Moon dances around the Earth, and the Earth around the Sun, eclipses occur at those precise moments when all three bodies are in alignment. During these brief and special events, the Sun and the Moon assume strange and unsettling aspects – the stuff of myths and legends throughout human history.

Nowadays, our ancestral fears have been allayed by our knowledge of celestial mechanics and our understanding of these events, and eclipses have become mass-audience spectacles, attracting hundreds of thousands or even millions to sites in the privileged path of the Moon's shadow.

What happens during an eclipse? Where and how should it be observed? How can it be photographed?

Paradoxically, eclipses of the Sun allow astrophysicists to lift a little of the veil concealing the Sun's secrets. These phenomena continually add to our knowledge of solar physics, and the relationships between Earth, Moon and Sun, since the Sun and the Moon present to us, during eclipses, aspects of their faces not normally seen. Total eclipses of the Sun have revealed the solar wind – a tenuous plasma atmosphere continually streaming out into the space between the planets and bathing the Earth's magnetosphere.

The corona, the Sun's extensive outer envelope, still holds many secrets, and is nowadays intensely studied in its rôle as the main active source of the Sun's effect upon the Earth.

List of illustrations and tables

Chapter 5

Chapter 6

Appendix B

Appendix D

Appendix E

Addresses and bibliography

Tables

Plates

The plates appear between pages 132–133.

Plate 1

Fig. 1. An excellent synthetic photograph from a Japanese team observing the eclipse of 24 October 1995 from Rajasthan, India. With digital processing it is possible to amplify the subtle modulations of light and improve the visibility of slightly contrasted detail. (Prof. Hiei and colleagues, Mesai University.)

Plate 2

Fig. 1. The eclipse of 11 July 1991. Computer-processed composite image from several original images of exposure times from 1/60 to 4 seconds, using a 12.5-cm telescope of focal length 1 metre. (Mesai University.)

Fig. 2. Considerable activity on the Sun, with the ejection of plasma streamers, taken with the Lasco coronagraph on 21 August 1996. On this part of the image, above the coronal sheet at the plane of the equator, there appear ejections of ionised gases curling around magnetic field lines. This triple coronagraph can study both the inner corona (a few thousand kilometres above the surface) and the outer corona (about 20 million kilometres above the surface). The corona in this image is four million kilometres wide. (NASA/ESA.)

Fig. 3. The total eclipse of the Sun of 26 February 1998, seen here from Guadeloupe in the Caribbean. The corona was also observed from space by Yohkoh and SOHO for several hours around totality. On this montage Yohkoh's image of the corona replaces the Moon's disc, surrounded by the image of the white-light corona obtained by P. Martinez (Adagio, S.A.F.) near Anse-Bertrand, Guadeloupe, using a radial mask, at 18 h 32 UT. The exposure time was 2 seconds. (Montage by V. Koutchmy.)

Fig. 4. Small polar jets observed at soft-X-ray wavelengths, produced at about 3 million K. These jets are probably related to the polar plumes observed during total solar eclipses. This photograph was taken on 2 September 1995 with Yohkoh's X-ray telescope at a rate of one image every 32 seconds. (Japanese Institute of Space and Astronautical Science/Lockheed Palo Alto Research Laboratory.)

Plate 3

Fig. 1. Three types of solar eclipse: a total eclipse, when Sun, Moon and Earth are aligned, and the Moon is close enough to the Earth to be able to cover the Sun completely; an annular eclipse, when Sun, Moon and Earth are aligned, but the Moon is further from the Earth and its diameter is not sufficient to cover the Sun completely; and a partial eclipse, when Sun, Moon and Earth are not closely aligned, and the shadow cone only partially reaches the Earth.

Fig. 2. The zodiacal light and comet Hale–Bopp, photographed on 25 March 1997, about one hour after sunset near the observatory at Saint-Michel-de-Provence. This 7-minute exposure was guided on the stars, using Kodak Ektar 1000 and a 20-mm wide-angle lens. Notice the comet's two tails. The white dust tail extends along the plane of the comet's orbit, and the blue ion tail is deflected by the solar wind. (P. Bourge.)

Fig. 3. Montage showing different phases of a total eclipse, taken in India during the eclipse of 16 February 1980. (Japanese Journal *Newton*; special edition of 1983.)

Plate 4

Fig. 1. Photograph taken in Chile on the shore of Lake Chungara during the total eclipse of 3 November 1994. This montage was created with an image-processing package. After digitisation, photographs of the partially eclipsed Sun were added to the scene. Note Venus at top right. (Y. Méziat, montage by P. Guillermier.)

Fig. 2. Accurate timing and recording of eclipses by civilisations in Mesopotamia, ancient China, Arabia and Europe soon led to the discovery of the 18-year saros cycle. Archive material on engraved bones, terracotta tablets and papyrus, giving dates and places for the passage of the Moon's shadow, led British researchers from Durham University and the Royal Greenwich Observatory to conclude that, in the last 2,700 years, the Earth's rotation has slowed by 0.0017 s per century. (P. Guillermier.)

Plate 5

Fig. 1. On 1 May 1996 a coronal mass ejection was recorded by the externally-occulted Lasco C3 coronagraph aboard SOHO. Note comet C/1996 B2 Hyakutake entering the field. These jets of coronal matter are moving at speeds of several hundred kilometres per second. This composite image was created by a process of subtraction, from images taken before and during the ejection. This improves the visibility of the ejections. (NASA/ESA.)

Fig. 2. CCD images of the east and west inner corona, during the eclipse of 11 July 1991 in Mexico, through an interference filter centred on the green line of FeXIV at 530.3 nm. (Y. Suematsu, NAO/Mitaka.)

Fig. 3. The eclipse corona of 30 June 1973, observed through a neutral-density radial filter. Notice the large streamers in the east–west equatorial region, the coronal hole to the north, and earthshine on the night side of the Moon. (IAP/CNRS.)

Fig. 4. A set of highly enlarged images (on a scale of only seconds of arc), showing the coronal plasmoid observed at the eastern limb during the eclipse of 11 July 1991 by the 3.6-m CFH Telescope on Mauna Kea, Hawaii. Each image is centred on the centre of gravity of the plasmoid, which cancels out its motion, reversing the motion of the more 'stationary' surrounding structures. An interference filter at about 637 nm, excluding any 'cool' lines, was placed in front of the CCD video camera at the telescope's prime focus for the 200-second observation recorded here. (IAP/CNRS.)

Plate 6

Fig. 1. The corona at the eclipse of 11 July 1991, observed through a neutral-density radial filter. Notice the great streamers extending, unusually, northwards and southwards. (J.-P. Zimmermann, IAP/CNRS.)

Fig. 2. Eclipse sequence, 25 October 1995, from Rajasthan, India. (P. Guillermier.)

Fig. 3. This picture shows the great twentieth-century French astrophysicist Bernard Lyot and his wife Madeleine, preparing spectrographs for the eclipse of 25 February 1952. (A. Dollfus, collection of G. Lyot.)

Fig. 4. On 10 May 1994, the evening Sun sets, ruddy and partially eclipsed, through clouds. (P. Guillermier.)

Plate 7

Fig. 1. The eclipse of the Moon on 3/4 April 1996. The Moon slowly emerges from the umbra in this photograph, taken at 0 h 40 UT on 4 April through a 180-mm Takahashi Mewlon telescope, of focal length 1,300 mm, with Fuji 400 ISO film and an exposure time of 5 seconds. (P. Guillermier.)

Fig. 2. Shuttle astronauts in orbit have a prime view of the influence of the Sun's activity on the Earth's atmosphere, as particles from the Sun penetrate the magnetospheric cavity. (NASA.)

Fig. 3. Sequence taken during the partial lunar eclipse of 8 October 1995. The rising Moon gradually leaves the Earth's penumbra. (P. Guillermier.)

Plate 8

Fig. 1. Photograph taken in India during the eclipse of 24 October 1995, through an 83-mm refractor of 1.2 m focal length, with parafocal radial filter. Exposure time 2 seconds, on Fuji NPS 160 film. (F. Diego, University College London.)

Fig. 2. Observation of the partial phases of an eclipse of the Sun requires the use of protective filters. Welder's goggles (No. 14) – as worn by these two observers during the eclipse of 3 November 1994 in Chile – can be very effective in reducing brightness and filtering out dangerous infrared and ultraviolet radiation. (Y. Méziat.)

Fig. 3. False-colour photographs of the Siberian eclipse of 22 July 1990. Images are processed in this way for better analysis of disturbed structures in the corona. (IAP/CNRS.)

Fig. 4. Various countries have commemorated, with postage stamps, the passage of the Moon's shadow across their soil. (Collection of L. Baldinelli.)

1

The Sun: our local star

Astrophysicists specialising in the study of stars tell us that the Sun is a medium-sized star, of spectral type G2, lying in the middle of the main sequence of stellar evolution on the Hertzsprung–Russell diagram. We now examine in detail what this means, and look at the Sun's radiation, its energy source, its internal and external structure, cyclical behaviour and future.

THE SUN AS A STAR

The Sun is just one of countless stars in our Galaxy. There is nothing special about its position or its intrinsic properties: lying two-thirds of the way out from the Galactic centre, it inhabits a spiral arm similar to those seen in other spiral galaxies in our Universe.

The Sun was born 4.7 billion years ago, a product of the collapse of an immense cloud of interstellar gas. It is now a mighty sphere of hot gases, mostly hydrogen and helium, 1,391,994 km across (nearly 110 times the diameter of the Earth). Its mass is 1,989 $\times$ 10^{30} kg (332,946 times that of the Earth). Its mean density is low, at 1.41 g cm^{-3} (compared with the Earth's density of 5.52 g cm^{-3}), indicating the presence in abundance of light elements.

The Sun's energy originates in nuclear reactions at its core, as hydrogen atoms fuse to become helium. The mean surface temperature is 5,770 K. The temperature at the nuclear core is thought to be at least 15 million K.

The distance from the Earth to the Sun varies, from 147.1 million km in January to 152.1 million km in July. In January the Earth is at a point in its orbit known as perihelion, and in July it is at the point known as aphelion. The mean Earth–Sun distance is used as a reference, and is known as the astronomical unit (1 AU = 149,597,900 km).

The Sun's axis of rotation is inclined at 89° 49′ from the plane of the ecliptic (the plane of the Earth's orbit) and its rotation period, measured using the Doppler effect, is about 25.4 Earth days. The Sun exhibits differential rotation; it spins fastest at its equator (24.6 days) and slowest at its poles (35 days). This is in part responsible

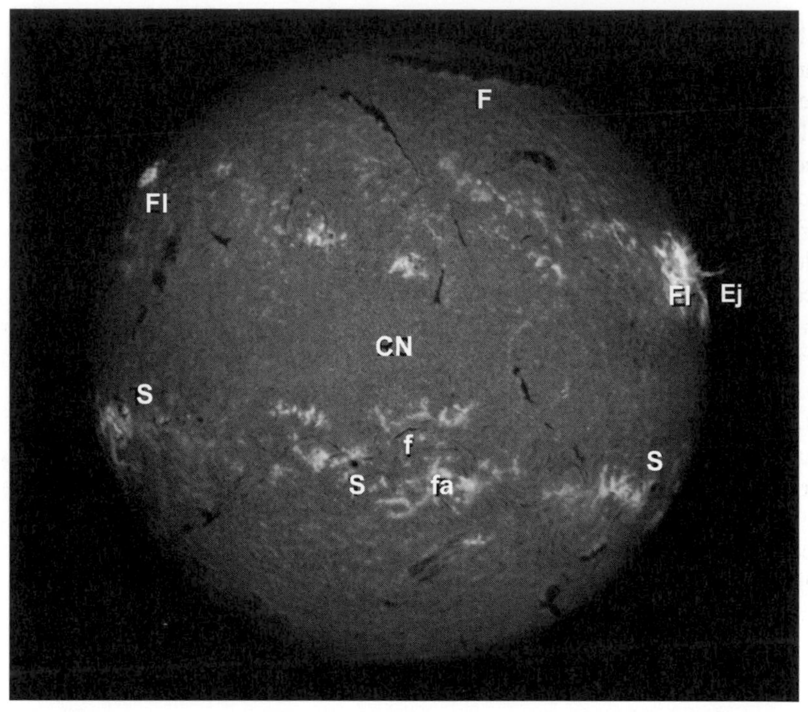

Fig. 1.1. The Sun at solar maximum, as seen in Hα light (hydrogen, Balmer series, $-\ \alpha$ line). S = sunspot; fa = facula; CN = chromospheric network; Fl = flare; Ej = surge ejection; f = plage filament; F = high-latitude filament. (Sacramento Peak Observatory/National Solar Observatory, USA.)

for the Sun's magnetic field, which is permanently active at its surface and is the cause of active solar phenomena.

SOLAR RADIATION

The Sun emits enormous quantities of energy, which can be measured by studying how much of it is received at the Earth's surface. This amount of energy, corrected for absorption in the Earth's atmosphere, is called the solar constant, S. The value of the solar constant S is about 1,400 W m^{-2}. Allowing for the distance from the Earth to the Sun, and assuming that the Sun radiates uniformly, it is estimated that each square metre of the Sun emits 63,000 kW – a total energy of 3.86×10^{26} W.

The Sun's light is emitted principally at the photosphere, in a range of wavelengths from 200 nm to 2,000 nm (1 nanometre = 10^{-9} m). The distribution of energy from ultraviolet to infrared corresponds closely to that of a black body at a temperature of 5,700 K. Maximum radiation is in the blue–green at 450 nm, and most of the energy is emitted in the visible spectrum between 400 and 700 nm. The low-resolution solar spectrum appears to be continuous, but greatly dispersed, and is crossed by a large number of lines caused by the selective absorption of certain wavelengths of chemical elements present in the solar atmosphere. This temperature and these spectral lines allow us to classify the Sun in its correct place as a member of the stellar family.

Observations have shown that stellar spectra themselves may be classified in a continuous series based on only one parameter: temperature. The spectral types, O, B, A, F, G, K, M, R, N, and S, which are subdivided decimally, are characterised by the colour of their continuous spectra and the absorption lines of their constituent chemical elements. Since the time of Hipparchus (second century BC), the brightness of stars has been arranged on a scale of *magnitudes*. This decimal scale is proportional to the logarithm of the star's luminous flux, and extends in both directions from zero. A star of magnitude 5 is 2.5 times brighter than one of magnitude 6, and so on. It seems likely that if a star is of high luminosity, its temperature must also be high. In 1913, a Danish astronomer, Ejnar Hertzsprung, and an American astronomer, Henry Norris Russell, published a diagram plotting absolute magnitudes; that is, the magnitudes of stars at a standard reference distance, against their spectral types. They showed that the majority of stars group themselves along a narrow band, known as the main sequence, with luminosity decreasing from type O towards type M.

The spectral types, with their corresponding temperatures, and the principal constituent elements and colours of the stars, are shown in Table 1.1.

THE INTERNAL STRUCTURE OF THE SUN

Let us now examine the architecture of the Sun's globe.

Matter is in hydrostatic equilibrium at all points in the Sun, as gravity, pulling its gases towards the centre, is counteracted by forces of pressure which prevent it from collapsing inwards. These latter arise from the gases and from the pressure of radiation proceeding from the core outwards. (Appendix describes the range of reactions governing the workings of the Sun.)

The Sun's internal structure may be divided into three zones:

- The core, of radius about 200,000 km (one third of the Sun's radius), where the nuclear reactions occur. Temperature here varies from about 15 million K at the centre to 7 million K at the periphery of the core.
- The radiative zone, extending from 0.3 to 0.7 of the solar radius, where the temperature is no longer high enough to support nuclear reactions. Energy is transported outwards by radiation, following a process of emission, absorption, re-emission and re-absorption of photons. The temperature at the exterior of the zone is 2 million K.
- The convective zone, extending from 0.7 of the solar radius out to the surface. Here, mighty convection flows ensure a more effective transfer of energy towards the exterior. This convection may be observed as granulation on the Sun's visible surface. Convection, together with the Sun's rotation, gives rise

Table 1.1. Stellar spectral types

Spectral type	Temperature (K)	Principal elements	Colour (spectral features)
O	25,000 and above	Highly ionized elements: helium II, silicon IV Hydrogen lines	Blue
B	15,000–20,000	Neutral helium Hydrogen lines	
A	10,000–15,000	Hydrogen lines H and K lines of ionized calcium present	White
F	6,000–8,000	H and K lines of ionized calcium Stronger lines of ionized metals: titanium II, iron II Hydrogen lines	
G	6,000	H and K lines of ionized calcium Metallic lines of calcium I, iron I Hydrogen lines Molecular CH and CN bands present	Yellow
K	4,000	H and K lines of ionized calcium Strong lines of neutral metals Strong bands of molecular CH Lines of ionized metals and hydrogen	
M	2,000–3,500	Molecular and titanium oxide bands	Red
R	3,000	As M, with proportionally less oxygen	
N	2,000–2,500	Carbon as molecular C_2 and CN	
S		Oxidized carbon, CO Zirconium oxide present	

to the solar magnetic field and to those phenomena characteristic of solar activity: sunspots, faculae, filaments, flares, prominences and coronal holes. The polarity of the magnetic field is reversed every eleven years, which suggests that a dynamo mechanism is involved. This mechanism is still only partially understood, but is thought to have some involvement with the convective zone.

These ideas are based on theoretical models. The science of helioseismology explores the Sun's internal structure. By measuring the velocities of vertical motions using the Doppler effect, helioseismologists can 'listen' to the vibrations of the Sun. In this way, sound waves emanating from the solar interior tell us of conditions in the interior. The expansions and contractions of the solar atmosphere produced by these vibrations stem from displacements of matter occurring in 'pressure modes'. These vertical movements do not exceed 10 metres in amplitude, and the vertical velocities are of the order of 10 to 20 cm s^{-1} for each mode. These velocities may be detected

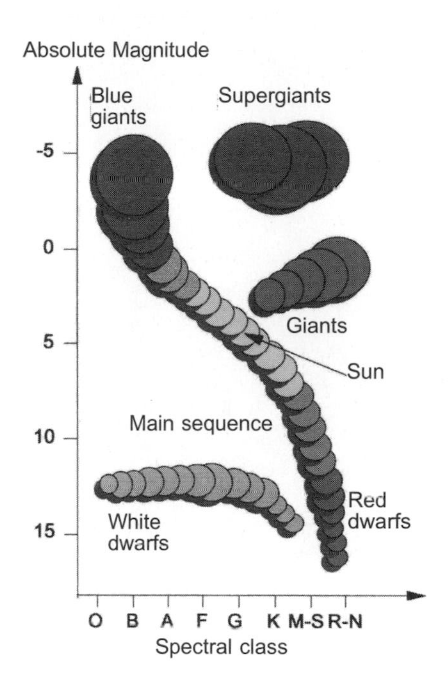

Fig. 1.2. A simplified Hertzsprung–Russell diagram. The curving line across the centre represents all those stars, of whatever age, which are still at the stage of hydrogen-burning at their cores. The Sun, of spectral type G2 and absolute magnitude 4.86, lies towards the middle of this so-called main sequence. At the top of the curve lie hot, blue stars of types O and B, and at the bottom the red dwarfs of type M. In the other parts of the diagram are found those stars which have exhausted their reserves of hydrogen. At the top, having left the main sequence, are giants and supergiants, in which nuclear reactions still proceed. However, they are burning other elements such as helium, carbon and oxygen. These stars are of great intrinsic brightness, which varies little with colour. Below the main sequence are the white dwarfs: rare, hot stars of high absolute magnitude.

as tiny spectral shifts in the Sun's light. The identification and analysis of these vibrations can be the source of important information, since the speed of propagation of sound varies according to the density and temperature of the medium through which it travels. The Sun's signal, broken down into a multitude of frequencies and amplitudes, proves that it is not a perfectly homogeneous sphere of gas.

Pressures and temperatures prevailing at the Sun's centre may be calculated on the assumption that its hydrogen–helium mixture obeys ideal gas laws: the density at the Sun's core attains 151 g cm^{-3}, and the pressure 233×10^9 bar – 233 billion times atmospheric pressure at the Earth's surface. This implies a core temperature of about 15 million K. Such a temperature supports nuclear reactions, and the totally ionised

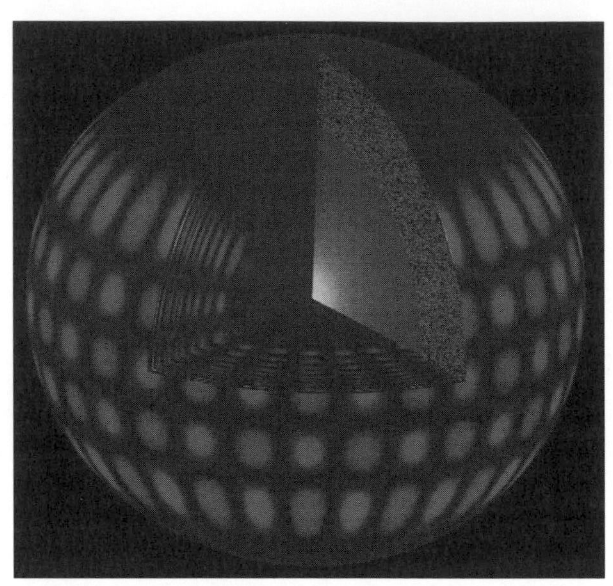

Fig. 1.3. Helioseismology has revealed and measured the various modes of acoustic waves resonating within the atmosphere of our star. In this simulation, light and dark zones correspond to strong amplitudes in the vibrations. The Sun is seen here in oscillatory mode, with precise values for radial, angular and rotational wavelengths. Analysis of these data adds to our knowledge of the Sun's internal structure. (NSO.)

matter exists here as a plasma of electrons and atomic nuclei. Photon energy emitted at the centre of the Sun is not directly transmitted to the surface, but undergoes a succession of absorptions and emissions, interacting with atoms in the solar gases, as it moves outwards. This radiative transfer causes an extension of the energy spectrum of photons, gamma-rays, X-rays and ultraviolet radiation into longer and more visible wavelengths. Calculations have shown that a photon of a given wavelength spends two million years on its journey from the core to the surface. Compare this with the two-second journey of a neutrino, whose interaction with matter is negligible.

At the surface of certain massive stars, there has been detected a superabundance of elements created at their cores: for example, helium, nitrogen and even sodium. How can this transfer be explained without recourse to some mixing mechanism in the central regions, a hypothesis put forward to account for the deficit in solar neutrinos? Current studies are aimed at identifying the cause of matter transfer within the radiative zone. This fundamental point must be explained in order to ensure the best-quality models, since evolution in stars has much to do with the composition of their internal regions.

The Sun's chemical composition

Table 1.2 shows the elements present in the Sun, and their relative abundance in numbers of atoms or corresponding ions. Note the deficiency in lithium, beryllium and boron. For every billion hydrogen atoms/protons, only one atom of lithium is

Table 1.2. The chemical composition of the Sun

Element	Atomic number	Symbol	Relative abundance
Hydrogen	1	H	10^9
Helium	2	He	80×10^6
Lithium	3	Li	1
Beryllium	4	Be	0.3
Boron	5	B	0.7
Carbon	6	C	300,000
Nitrogen	7	N	100,000
Oxygen	8	O	700,000
Fluorine	9	F	250
Neon	10	Ne	300,000
Sodium	11	Na	1,600
Magnesium	12	Mg	30,000
Aluminium	13	Al	1,700
Silicon	14	Si	30,000
Phosphorus	15	P	350
Sulphur	16	S	16,000
Chlorine	17	Cl	250
Argon	18	Ar	4,000
Potassium	19	K	80
Calcium	20	Ca	1,600
Scandium	21	Sc	1
Titanium	22	Ti	70
Vanadium	23	V	6
Chromium	24	Cr	250
Manganese	25	Mn	120
Iron	26	Fe	8,000
Cobalt	27	Co	50
Nickel	28	Ni	850
Copper	29	Cu	45
Zinc	30	Zn	26

found. Relative abundance is greater for those elements with even atomic number (that is, more stable) than for those with an odd number of protons. A relative peak in abundances is noted for heavy elements like iron and nickel.

THE PHOTOSPHERE

The photosphere (Greek *photos/phos*, light, and *sphaira*, sphere) is the layer representing the outer visible surface of the Sun. The opacity of this layer, and hence its visible nature, stems from the presence of neutral hydrogen, negative hydrogen ions and energy transitions of free electrons.

The photosphere is the region of the Sun whose chemical composition can be most accurately determined, as it gives rise to several million identifiable absorption (Fraunhofer) lines. These atomic or molecular lines have absolute intensities from which the density of the absorbing atoms can be deduced, as well as the abundance of the element in question. The abundance of heavy elements in a star reflects the history of their creation by nucleosynthesis. Most of these elements are present on Earth. Convection causes the appearance at the photosphere of large numbers of turbulent cells of gas, which exhibit a reticular structure as seen from Earth. These bright cells, discovered by James Short in 1748, are called granules, or sometimes 'rice grains'. They are the sites of vertical ascending movements of masses of gas, at velocities of about 2 km per second. This convective motion sustains sound waves, which do not propagate into the regions above (the chromosphere and the corona), but only where the magnetic field allows. In the photosphere the waves are stationary. The granules are 300–2,000 km across, and they have typical lifetimes of a few minutes. The total number present on the solar surface at any time is about five million. This granular appearance is due to spatial temperature fluctuations: the granules are on average several hundred degrees hotter than the surrounding medium, and the different temperatures create differences in brightness. The photosphere is perfectly stratified and in hydrostatic equilibrium except for these convective fluctuations, which, although small, are sufficient to maintain the magnetism responsible for the Sun's activity.

At a higher level appear spicules, which are associated with strong magnetic fields (1,500 gauss at the photospheric level). These fields are very localised. The spicules

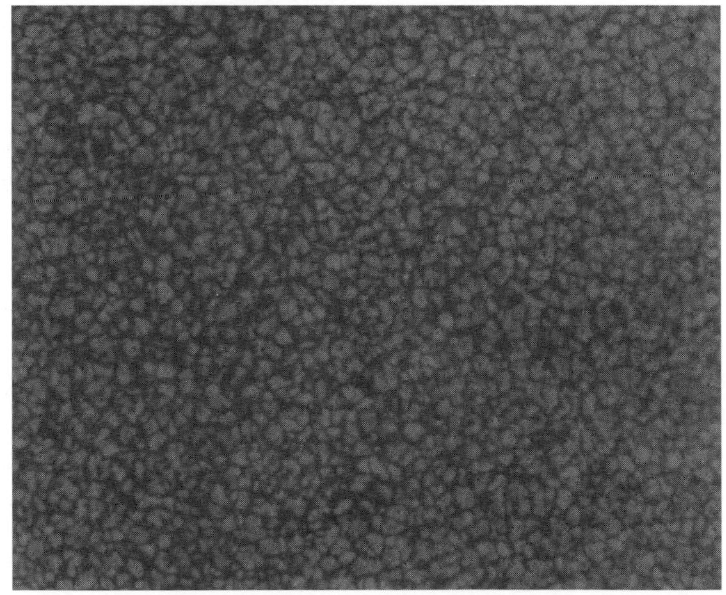

Fig. 1.4. This high-resolution image shows the photospheric gas cells causing granulation. The 'rice grains' are evidence of thermal convection within our star. (Pic-du-Midi Observatory/OMP.)

are part of the chromosphere, and are observable in Hα light at a wavelength around 656 nm. They are visible at the edge of the Sun, resembling little dark spikes in constant motion. Their average inclination is 20–40° from the vertical, their average diameter is 400 km, and they can reach heights of 10,000 km.

General motions and brightness variations affect the photosphere over areas of thousands of kilometres, with, more especially, oscillations in pressure. Mesogranulation, discovered by L. November in 1970, acts over distances of about 5,000 km, such disturbances lasting for approximately three hours.

In 1960, the phenomenon of supergranulation was revealed by R.B. Leighton while measuring horizontal velocity fields in these cells. This network had been known about for some time in the chromosphere. Supergranules are 30 times bigger than 'rice grains', and have a typical lifetime of 20 hours. Theorists are working towards an explanation of the existence of supergranulation cells and the oscillations affecting the Sun's surface. Current work suggests that evacuation of energy by the Sun is far more chaotic and irregular than previous models implied. It must, however, be remembered that the theory of radiative transfer in the stratified layers of the photosphere explains perfectly the phenomenon of darkening towards the solar limb, in an atmosphere where temperature decreases regularly towards the exterior.

THE CHROMOSPHERE

Beyond the photosphere, towards an altitude of about 500 km, temperature continues to fall until a minimum of 4,200 K is reached. From here on, the temperature begins to rise with increasing altitude, into the chromosphere. When looked at in strict thermodynamical terms, this phenomenon seems inexplicable, since the only source of heat is the Sun. From 5,000 K at its base, the temperature in

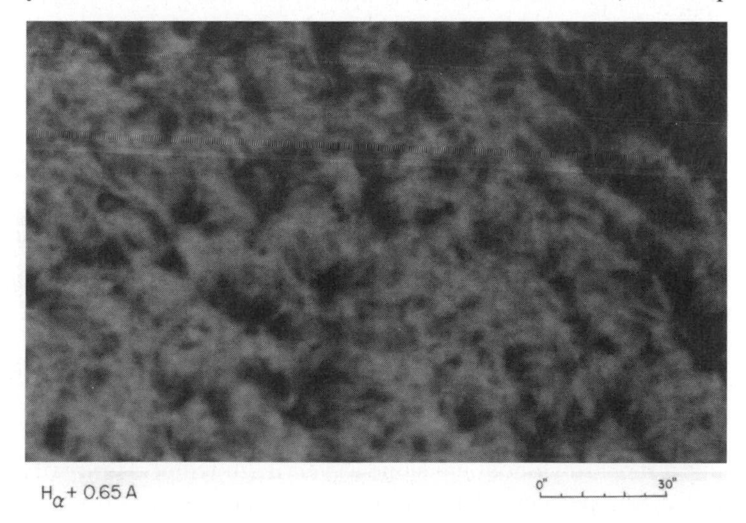

Fig. 1.5. The chromospheric network observed in the vicinity of the Hα line, in a calm region. (SPO/NSO.)

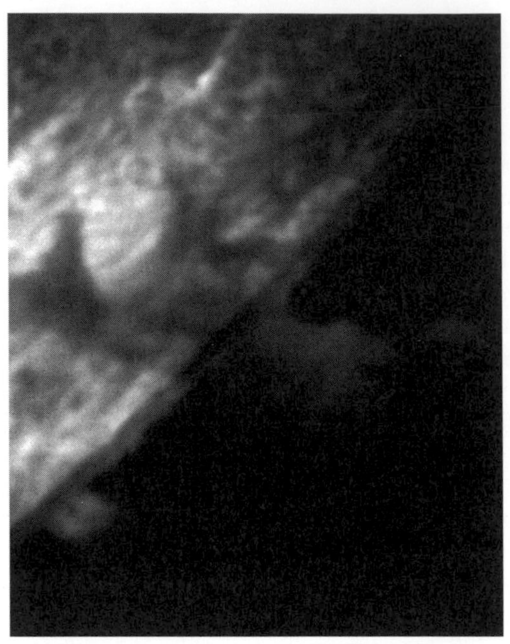

Fig. 1.6. The Sun's chromospheric fringe, with a small prominence at the edge of the solar disk, seen in Hα. (SPO-NSO.)

the chromosphere attains 9,000 K at 2,000 km and 20,000 K in its uppermost regions. Some heating mechanism due to the dissipation of magnetosonic waves might explain this rise in temperature.

The chromosphere (Greek *chromos*, colour) is so called because it may be seen as a fine coloured edge around the Sun during total eclipses. The striking pink hue has its origin in the most intense of all the chromospheric lines, that of Hα, mixed with a hint of orange from the D3 linc of helium and blue from other hydrogen lines.

In contrast to the inner photosphere, where temperature falls with altitude, the magnetic field apparently produces extra heating in the outer photosphere and chromosphere, and the temperature rises. This increase in temperature and turbulence in the chromosphere is linked with the development of plages, or faculae, which may cover several billion square kilometres. Elsewhere, events involving accelerated velocities occur, as in the spicules, with gases suddenly ejected upwards to redescend slowly into the inner chromosphere.

THE CORONA

The corona is a rarefied gaseous envelope at very high temperature, with densities of less than 10^9 atoms per cm^3. The luminosity of the corona falls off rapidly with increasing distance from the Sun. Magnetic forces make the structure of the corona very complex. In spite of the rarefied nature of its gases, collisions between high-velocity atoms lead to their ionisation. Ionised atoms and particles, such as electrons and protons, move around magnetic field lines. Intense ultraviolet and X-ray emissions, and emissions from strongly ionised atoms, raise temperatures to 1,000,000 K, and occasionally to 10 million or more. In the 1940s, Bengt Edlén and his co-workers, using spark discharge spectroscopy, identified the green line of FeXIV at 530.3 nm, and the red line of FeX at 637.5 nm. The identification of these spectral lines, which had long been observed in the corona during eclipses, was the basis for final confirmation of the enormous temperatures in the corona, together with a new kind of solar atmospheric physics based on the properties of ionised gases or plasma.

While temperatures lower down in the photosphere reach a minimum of 4,200 K, the corona may be at several million degrees. The phenomena involved in the generation of such temperatures are to this day not understood, and present a challenge for astrophysicists. Creating models of acceleration processes for different particles at sites in various layers of the solar atmosphere is a very complex business, because of the presence of the Sun's magnetic field and gravity. Mass-charge relationships in particles seem to play a part.

It has long been thought that this heating mechanism works through dissipation of shock waves, emanating from the base of the solar atmosphere as a result of turbulence in the convective region. Photospheric granulation is indicative of this turbulence, but nobody has yet shown how mechanical energy is transferred from the

Fig. 1.7. The total eclipse of 30 June 1973, from Loiyengalani, Kenya. This excellent photograph of the corona was taken through a red filter with a neutral-density radial gradient, which attenuates the inner corona, so that the whole corona can be photographed in one image. (High Altitude Observatory/National Center for Atmospheric Research, sponsored by the National Science Foundation, USA.)

photosphere into the corona. It seems certain, however, that microstructures in the magnetic field, visible in the chromosphere, have a decisive rôle in coronal heating. We are beginning to see models explaining this energy transfer. Studying the heating and acceleration of the solar wind within the corona is made difficult, however, by coronal holes, apparently essential to the fast solar wind, which defy all observational investigation at coronal level because of the rarefied nature of the plasma. The Swedish physicist H. Alfvén showed that the magnetic field creates magnetic or magnetohydrodynamic waves capable of modifying the mode of propagation of acoustic waves. Alfvén waves may also dissipate in the upper atmosphere of the Sun, accelerating it. An understanding of heating in the solar atmosphere may be found in studies of the dissipation of currents and the propagation of energy within small-scale magnetic structures.

On examining in detail the structure of the optical solar corona, we distinguish several components through the nature of their radiation:

- The emission corona, characterized by a large number of intense monochromatic lines. Here we can identify atoms of iron, chromium, nickel, calcium, magnesium, silicon and oxygen, all very strongly ionized;
- the diffuse corona, itself divisible into two components:
  - the K component, which scatters peripheral light and is the dominant factor for a distance of about one solar radius. In the K corona (so called from the German word *Kontinuum*) spectral lines are considerably broadened and may trespass into neighbouring lines, as a result of the high temperature of very rapid electrons, to the extent that these lines may no longer be seen, as they merge into the continuous spectrum;
  - the F component (F for Fraunhofer, discoverer of the lines of the solar spectrum), which prevails further out, scattering the Sun's light. Fine, cool dust, far from the Sun, ensures that scattering occurs without broadening the spectral lines, and they therefore do not merge with their neighbours.

Photographs of the corona around the disk during total eclipses of the Sun, or on the disk in X-ray images, taken by satellites, show its extreme heterogeneity and geometrical variability.

The main structural peculiarities of the solar corona are distinguished by holes and coronal streamers.

Coronal holes, which show up in X-ray photographs as dark patches, tend to be localised near the poles when small, but may stretch from one pole to the other along meridians. They are regions where the coronal magnetic field is unipolar, and where the lines of force are open to outer space. These holes are seen only in the corona, and do not exist in the photosphere. Images taken during total eclipses show zones where matter is drawn into long, divergent radial filaments, which are slightly curved. These fine structures are called polar plumes. Coronal holes are the coolest and least dense regions of the corona. They are relatively stable over time and turn with the Sun, maintaining their integrity as it rotates.

Coronal streamers occur at all latitudes. They are associated with bipolar photospheric regions, where magnetic field lines close up to form a loop. These

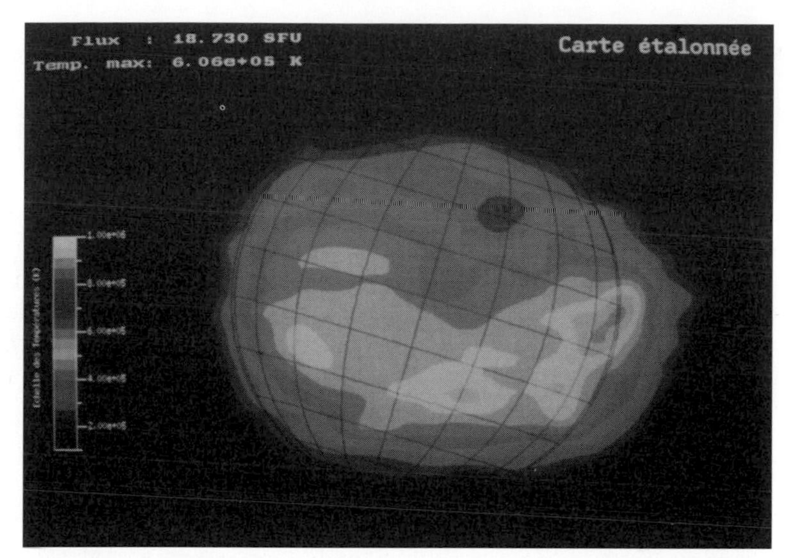

Fig. 1.8. Radio image of the inner solar corona at 408 MHz, from the Nançay Observatory, 29 May 1986. A routine observation, showing coronal holes and high equatorial activity. (Paris Observatory.)

streamers have wide bases, and form immense arches over complex and very delicate chromospheric structures. The streamers taper with increasing distance from the surface, appearing as very long luminous filaments, extending outwards for several solar radii.

Table 1.3, derived from various sources (VLA model, Allen, J.M.Fontenla and associates), shows the characteristics of a mean model of the quiet solar atmosphere. Altitude, moving outwards from the Sun, is shown in positive figures, 0 indicating the solar surface; that is, the region from which optical radiation essentially emanates. By definition, the surface of the Sun corresponds to the level where optical density $\tau 5$ at 500 nm is equal to unity.

It should be noted that densities fall off with altitude for various reasons: firstly, because the pressure gradient balances the force of gravity acting on the gas (hydrostatic equilibrium); secondly, because this pressure is modified by temperature variation; and thirdly, because of magnetic forces, which are dominant in the inner corona. Temperature passes through a minimum at about 500 km, thereafter increasing extremely rapidly with altitude.

Appendix B includes a more extensive description of the kind of coronal physics made possible during eclipses. A good introduction to the analysis of the problem of radiative transfer in the atmosphere of the Sun and other stars can be found in *Astrophysique Générale* by J.-C. Pecker and E. Schatzman and in the classic book of A. Unsöld (see Bibliography.)

Table 1.3. Characteristics of a mean model of the quiet solar atmosphere.

h	τ_5	T	v_t	n_H	n_p	n_e	ρ
1,000,000	0	1,000,000	20	1.0E+06	1.0E+06	1.0E+06	2.4E–17
100,000	0	1,500,000	25	1.0E+08	1.0E+08	1.0E+08	2.4E-16
10,000	0	1,000,000	30	3.0E+08	3.0E+08	3.0E+08	0.8E–15
2,218	0	100,000	11.7	5.57E+09	5.575E+09	6.665E+09	1.31E–14
2,200	1.21E–08	17,900	9.1	3.00E+10	2.483E+10	3.049E+10	7.05E–14
2,000	5.41E–07	8,400	7.2	1.203E+11	3.921E+10	4.313E+10	2.82E–13
1,200	5.19E–06	6,230	3.5	1.002E+13	8.783E+10	9.083E+10	2.35E–11
450	8.51E–04	4,460	0.65	4.192E+15	3.600E+10	4.714E+11	6.79E–09
150	9.92E–02	5,150	1.00	5.062E+16	3.579E+11	6.153E+12	1.19E–07
50	4.31E–01	5,790	1.40	9.558E+16	7.614E+12	1.980E+13	2.24E–07
0	1.00E+00	6,520	1.60	1.82E+17	6.014E+13	7.697E+13	2.77E–07
–50	4.13E+00	7,900	1.75	1.295E+17	6.668E+14	6.923E+14	3.04E–07
–100	2.36E+01	9,400	1.83	1.351E+17	3.826E+15	3.867E+15	3.17E–07

h : altitude in km
τ_5 : optical thickness at wavelength 500 nm (by definition, $I/I_0 = e^{-\tau}$, where I is emergent intensity and I_0 falling intensity from deep layers)
T : mean temperature in K
v_t : microturbulence velocity in km s^{-1}
n_H : total mean density (number of hydrogen atoms per cm^3) of neutral and ionised atoms
n_p : density of protons (number of protons per cm^3)
n_e : density of electrons (number of electrons per cm^3)
ρ : density in g cm^{-3}

SOLAR ACTIVITY

It is now time to look at solar activity in the photosphere and the chromosphere. To begin with, the photosphere harbours a very obvious manifestation of solar activity, which has long been known about: sunspots.

Sunspots may sometimes be visible to the naked eye. The Chinese had already observed this phenomenon a very long time ago, but with the invention of the telescope at the beginning of the seventeenth century, systematic study became possible. At that time, Galileo, Johann Fabricius and Christoph Scheiner discovered and studied the Sun's rotation, and later, its differential rotation, by observing sunspots.

Beginning in 1826, a German amateur astronomer, Heinrich Schwabe, carefully sketched the Sun's surface every clear day, noting all the spots, in an attempt to discover a hypothetical planet between the Sun and Mercury. He realised that this planet, if it existed, would be visible during transits across the face of the Sun. Schwabe also assembled an archive of 200 years of solar observations. This unprecedented mass of data led to the discovery, not of a new planet, but of an order concealed beneath the apparent chaos surrounding the emergence of sunspots. In 1843, Schwabe announced his discovery of a sunspot cycle with a frequency of ten years.

Swiss astronomer Rudolf Wolf, at the time director of Berne Observatory,

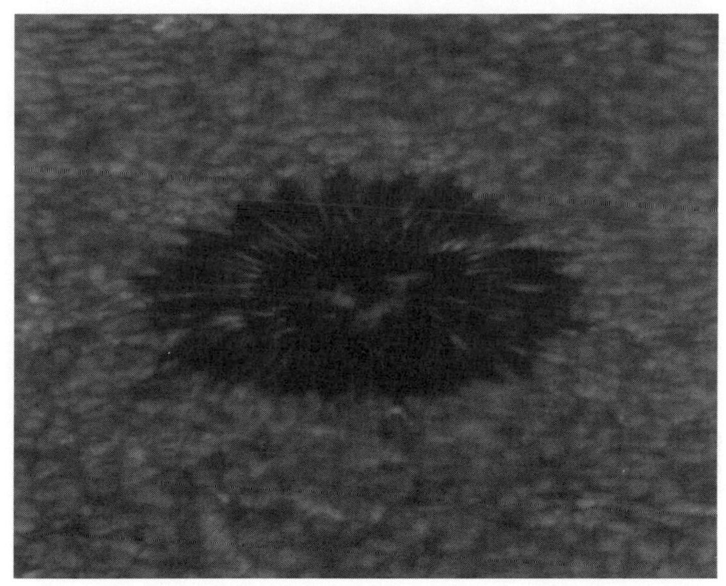

Fig. 1.9. Around the zone of the umbra in this composite image of a sunspot, where the umbra is superimposed as a negative image, the filaments of the penumbra, oriented tangentially to the solar surface, appear sharply defined due to the fact that their temperature is lower than that of the photosphere. The black spots in the umbra are in reality very bright. (SPO/NSO.)

confirmed Schwabe's cycle through systematic observations, refining it to a period of eleven years. Wolf worked out over time an arbitrary parameter R, known as the Wolf or Zürich Number, with the formula:

$$R = k(10g + f)$$

where k is a constant of normalisation depending on various parameters – the observer, image quality and type of instrument used to observe; g is the number of spot groups seen; and f is the total number of individual spots seen.

Sunspots have their origin in one of the fundamental phenomena of the Sun: the birth, evolution and disappearance of active centres around tubes of magnetic flux emerging from very deep down, perhaps at the base of the convective zone about 200,000 km below the surface. More localised solar magnetic fields are at the root of these disturbances. It should be noted that, although the general magnetic field of the Sun is weak (0 ± 1 gauss on average around the globe, or 2 ± 1 gauss per hemisphere, with different sign for the mean field of each hemisphere), fields around the spots or even around faculae are extremely intense (1,000–4,000 gauss). The Sun's differential rotation – more rapid at the equator than at the poles – perturbs the flux tubes which emerge from the photosphere in one zone of polarity, for example positive, and loop over through the chromosphere to fall back some distance away into a zone of negative polarity.

Observations confirm the magnetic nature of sunspots: spectral lines of metals are

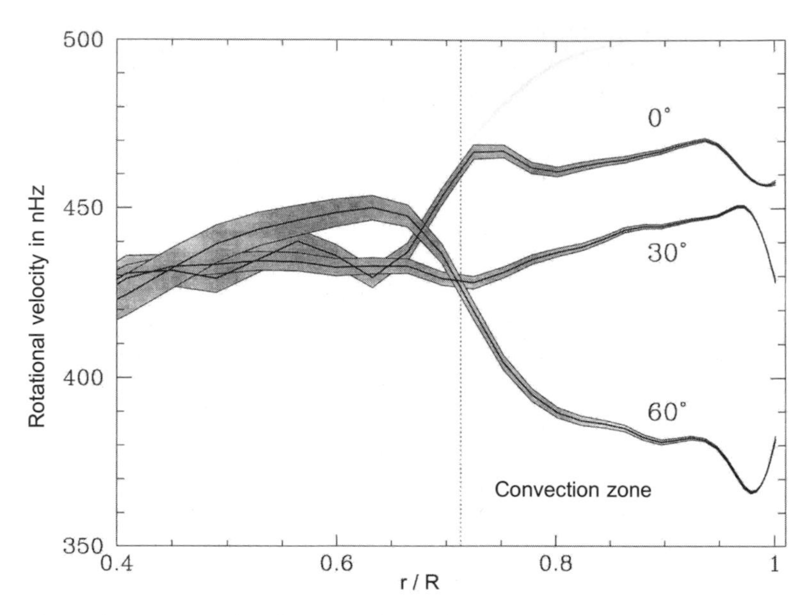

Fig. 1.10. Diagram showing the internal rotation of regions of the Sun as a function of depth and latitude, from data gathered by the MDI-SOI experiment aboard SOHO. Note that, at the surface ($r/R = 1$), the equatorial regions rotate faster. (ESA/NASA.)

split into several polarised components, in general the three lines of the Zeeman triplet. Measurements of the displacement of the lines, and of the amount of polarisation, suggest a very high value for the local magnetic field: about 3,000 gauss. The mechanism for the formation of spots explains the fact that they tend to concentrate within two bands on either side of the equator, between 5° and 45° N and S. In 1912, George Ellery Hale, studying the Zeeman effect, determined the magnetic origin of the spots by means of thc spectroheliosope. The Hale–Nicholson Law shows that the polarity of the spots is reversed from one cycle to the next. In 1938, Waldmeier classified sunspots into different categories in order to facilitate observations. Within the tube of magnetic force above the sunspots, gas pressure and density are weaker than in the more homogeneous medium of the photosphere. The temperature here is about 4,000 K, which is 2,000 degrees cooler than the photosphere. The nuclei of sunspots are of spectral type K and, although they are reddish in colour, to observers they appear black in contrast to the much hotter photosphere. Temperatures within spots are low enough to allow the formation of numerous diatomic molecules; for example, titanium oxide, magnesium oxide and calcium hydride. Even water vapour has been detected.

The nucleus of a sunspot, known as the umbra, is usually surrounded by an irregular area called the penumbra. Within the penumbra are seen alternating light and dark filaments about 500 km across. Their orientation suggests that magnetic fields in the penumbra are much more tangential to the solar surface, while fields in the umbra tend to be at right angles to it. Gases in the upper atmosphere of the Sun flow in these magnetic fields at velocities of about 4 km per second, while lower down, they escape at 2 km per second. This phenomenon – discovered by John Evershed in 1909 – remains unexplained.

Spot groups, no matter how complex, are usually bipolar. The leading spot has opposite magnetic polarity to the following one. Sunspots may be between 2,000 and 50,000 km in diameter, and their numbers vary from zero at times of minimal solar activity to over 200 at solar maximum.

In the chromosphere and the corona, solar activity manifests itself in faculae, prominences, and flares. Prominences are condensations of partially neutral cool coronal plasma. They assume the shapes of their local magnetic field, which produces the necessary force to sustain them in the corona. Seen above the solar surface in Hα light, they are denser and cooler than their surroundings, and contain greater numbers of hydrogen atoms, which scatter in all directions and strongly absorb light from the photosphere and part of the chromosphere below them. They therefore show up as dark filaments on the disk, but appear very bright against the background of the sky when seen protruding from the solar limb, as a result of the great amount of scattered light, or, when observed in the extreme ultraviolet (EUV), as a result of de-excitation.

There are various types of prominence:

- Quiescent prominences, which are relatively stable and of modest heights, below 100,000 km. These are long-lived, and they may survive for several solar rotations.
- Active prominences, which differ from the former by virtue of their rapid internal motions. They may last from a few minutes up to some tens of hours, and their heights range from 100,000 to 500,000 km.

Fig. 1.11. A large prominence observed in Hα light. (SPO/NSO.)

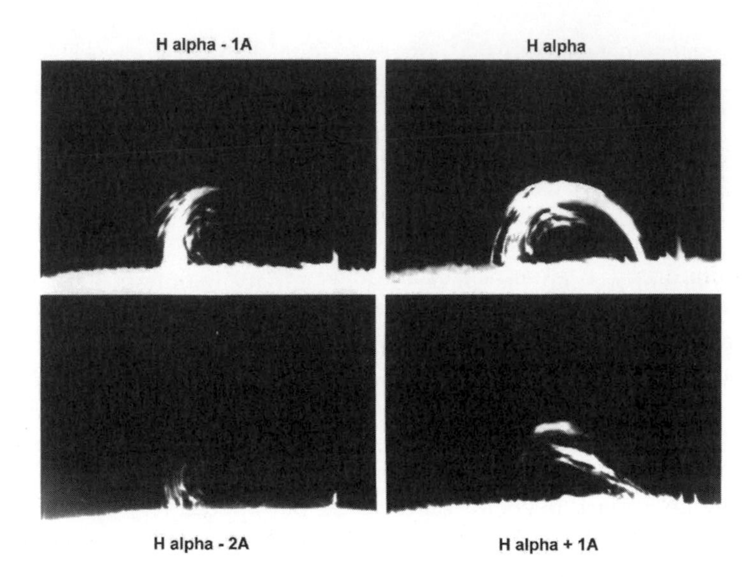

Fig. 1.12. Loop prominences at the solar limb, seen through a narrow Hα filter set at different positions relative to the line. High velocities, producing Doppler effects, are apparent. (SPO/NSO.)

- Eruptive prominences, rushing vertically upwards at great speeds (100–1,000 km s^{-1}), sometimes to altitudes of over a million kilometres. These events may, in spite of the considerable gravity, result in amounts of matter escaping into space.
- Surge and spray, forming a vertical pillar above the chromosphere. This phenomenon, linked to chromospheric eruptions, sees gases rising at several hundred kilometres per second to heights of tens of millions of kilometres. Rising sections may appear to tear themselves away from the magnetic structure.

As active regions, with their attendant sunspots, evolve, configurations of the magnetic field resulting from the superposition of bipolar or single centres of activity may appear, and give rise to solar flares. Lesser examples may affect the surface over several hundred million square kilometres, but the largest may affect up to five billion square kilometres. During these eruptions, magnetic field energy is converted into electromagnetic and kinetic energy, and temperatures of 20–100 million degrees may be attained. Non-thermal effects include the production of gusts of charged particles (electrons and protons) along lines of magnetic force. These particles may remain trapped in the vicinity of the Sun by closed magnetic field lines, or, if the lines are 'open' or detached, they may escape into space, some reaching the Earth. These may cause disturbances in our upper atmosphere, interfering with radio waves and triggering aurorae. Seen in Hα light, a solar flare is characterised by the appearance of very bright filaments on the solar surface. At the limb, loops may be seen rising to several tens of thousands of kilometres within the inner corona. Energetic coronal phenomena precede chromospheric flares by several tens of minutes, but, although seemingly fundamental, they cannot as yet be explained satisfactorily.

SOLAR CYCLES

Solar activity, of which sunspots are one of the most obvious manifestations, does not proceed uniformly. It varies according to a periodicity of 11 years – an interval called the Schwabe cycle, after its discoverer. At solar minimum, few spots are observed. Then, as the cycle continues, the number of spots increases regularly, and groups become ever more complex while spots become larger. At solar maximum, the greatest numbers of spots are seen, whereafter numbers begin to decrease towards minimum.

At the beginning of the cycle, at the time of solar minimum, spots appear at higher latitudes around 40°–42°, but as the cycle advances they appear closer to the equator, normally at 0°–10° in the months before the coming minimum. The large groups typical of solar maximum tend to occur at latitudes around 10°–25°. No spots are ever seen around the poles, but only in two zones lying symmetrically either side of the equator – the so-called 'royal zones', extending roughly from 45° N to 45° S.

Richard Carrington (1859) and Spörer (1874) were the first to point out the systematic drift in latitude of mean spot positions as the solar cycle progresses.

Spots have their own motions with reference to their surroundings. For example, while the Sun's equatorial region rotates in 24.6 days, a spot near the equator takes on average a little longer – 25.1 days – to make the same journey. The spot therefore

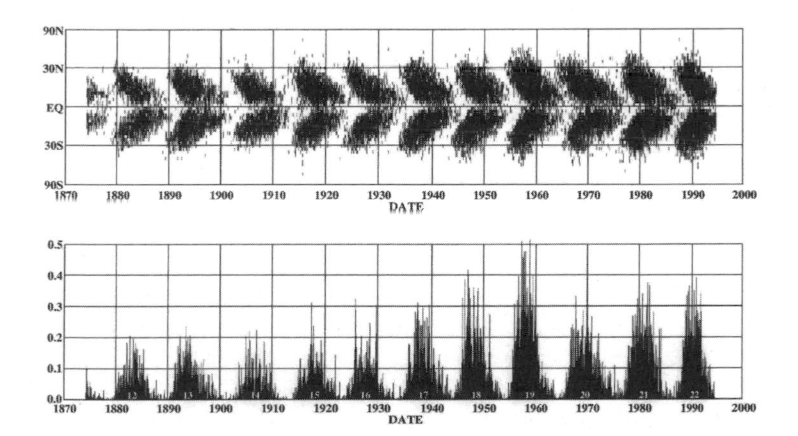

Fig. 1.13. During a cycle of activity, a plot of the changing positions at which spots appear either side of the equator produces the 'butterfly' effect. At solar minimum, spots eventually appear at higher latitudes, halfway between the poles and the equator, appearing in lower and lower latitudes as the cycle advances. On the chart, showing spot numbers over 120 years from 1875 to 1995, the 11-year cycle is apparent. Numbers alloted to cycles are indicated. (David Hathaway, NASA Marshall Space Flight Center.)

has a retrograde motion of some 3,600 km per day. If, at any given moment, in the northern hemisphere the leading spot has a south polarity, the follower will have opposite polarity. However, in the southern hemisphere the leader will be of north polarity and the follower of south polarity. Moreover, if, in the course of a cycle, leading spots in one hemisphere have a certain polarity, they will exhibit the opposite polarity during the next cycle in the same hemisphere. It seems that the real solar activity cycle lasts not for 11 years, but for 18 years. The Sun has a globally dipolar magnetic field, which reverses itself every 11 years – about three years before the appearance of the new cycle of spots. As the last spots of a cycle are still forming two or three years after the appearance of the first spots of the next, the total duration of a cycle may be said to be 18–22 years. Successive cycles therefore overlap, all the while maintaining their separate identities in latitude.

Over and above the Schwabe cycle, a supercycle or 'Gleissberg cycle' with a periodicity of 80–100 years, has been shown to exist.

Finally, there is the anomaly of the 'Maunder minimum', named after the English astronomer E. Walter Maunder. This occurred between 1650 and 1710. Solar activity was greatly reduced, being confined to one hemisphere and then only near the equator. A modern theory implying 'some strange attractor' has been suggested to describe this behaviour.

SUN–EARTH INTERACTIONS

With his upright stance, prehistoric man was the first creature on Earth to be able to lift his eyes towards the stars. The rhythms of his life were reflected in celestial events, and he could make connections between the Sun, the Moon and his own environment.

One of the ancient civilizations for which astronomy assumed great importance was that of the Egyptians. The rising and setting of certain stars was linked to numerous decisions, both administrative and agricultural. The Sun was worshipped in Egypt in its various forms: as Khepri, god of the rising Sun; Ra, of the burning midday; and Atum, the god moving into the night.

During the 1950s, scientists such as Donald Menzel pointed out that there were apparent links between times of sunspot maxima and minima, and meteorologically dependent items such as tree growth and wheat prices.

We can list the main interactions between Sun and Earth, starting with the most obvious.

Day and night

The rotation of the Earth fixes the rhythm of our days and nights. However, 900 million years ago, in Palaeozoic times, the rotation period was only 18 hours. The American planetologist Charles Sonett, of the University of Arizona, came to this conclusion while studying coastal sediments in Australia, and in Utah and Pennsylvania. Spring and neap tides leave their marks in sediments, and can be read by experts; and there is also an annual cycle linked to the relative positions of the Earth and the Sun, affecting tides at the equinoxes and solstices. It is therefore possible to identify a year in sediments, and, for each year, to count the number of neaps. The period of the Moon's orbit, and its distance, may be calculated from this number. Sonett has shown that the number of neaps decreases as we look back through the years, from evidence in sediments 900, 600 and 300 million years old. The conclusion is that the Moon has moved away from the Earth at a rate of about 3.25 cm per year. Over 900 million years this adds up to 29,000 km – approximately 7% of the present Earth–Moon distance.

The tides

The principal joint effect of the gravity of the Moon and the Sun upon the Earth – or more exactly, upon the fluid elements of our planet – is the tides. The basis of this interaction – the universal gravitation for which Newton determined the formula – means that any two points attract each other as a function of their masses and inversely as the square of the distance between them. Also, Earth and Moon revolve around their common centre of gravity, which is located within the Earth at about three-quarters of the distance from its centre. The waters of the ocean nearest the centre of gravity experience only a slight centrifugal force, but those nearer the Moon experience a greater gravitational attraction, and are raised up towards our satellite. Conversely, the water at the other side of the Earth, and therefore further from the Moon, experiences less attraction from the Moon, and the centrifugal force, given the greater distance from the centre of gravity, has the effect of raising the water in the direction away from the Moon.

The rising of the oceans, both beneath the Moon and on the opposite side, fits into an ellipse whose major axis turns to follow the Moon. While the Earth rotates in 24 hours, the Moon moves in the same direction by 13° as seen from Earth (360° in 29.53 days). So the Earth has to rotate 13° more for the Moon to be above the same point, which takes 51 minutes. So the period of the tides is on average 24 hours 51 minutes, explaining the changing times of high and low tides as the days go by.

It is not only the Moon which affects fluids: the Sun also plays its part. However, because of its much greater distance, its effect is only 5/11 that of the Moon, in spite of its great mass. The two attractions work together when Sun, Moon and Earth are aligned at new Moon and full Moon. At these times the sea rises and falls more markedly, and spring tides occur. Near the equinoxes, on 22 March and 22 September, these spring tides are greatest, since the Sun is then over the equator, and its action is reinforced. At these two times, tidal coefficients exceed 105 at least once, and usually exceed 110. When Sun, Earth and Moon form a right angle, at first or last quarter, the forces of attraction of the Sun and the Moon work against each other, and the rise and fall of the sea is more moderate. These are the neap tides.

Climate

For about 20 years now, global warming has been the subject of much discussion. The mean rise in temperature over a century has been of the order of 0.6–0.8° C. The timescale here corresponds to the period of industrialisation of the developed countries, and it is tempting to link this warming with human activity, as we emit ever-increasing quantities of greenhouse gases into the atmosphere. Correlations between human activity and climate are far from being firmly established. In fact, climate is the sum of many exchanges and influences, not least the amount of energy received by the Earth from outside. Given its rôle, for example, in the succession of glaciations and interglacial periods which mark the Quaternary era, it is obvious that the Sun influences variations in the terrestrial climate. The Sun has not always been at the same distance from the Earth, though, as the eccentricity of our planet's orbit and its axial direction in space change slowly through time. The combined effect is to cause fluctuations in the amount of solar energy received by the Earth over periods of thousands of years.

These variables affecting the Earth in its journey around the Sun, however, can hardly explain the more rapid variations which have characterised the last 1,000 years. Over shorter periods, variations in solar activity seem to be involved. The Schwabe cycle, as has already been stated, has an 11-year periodicity, but rhythms of longer period – for example, of the order of 80–100 years, or even longer – are recognised. It must be remembered that the Sun has a weak dipole magnetic field essentially aligned with its axis of rotation. The Sun's differential rotation distorts the lines of force of this global dipole, and another magnetic component is created, parallel to the equator. Active regions are formed, and sunspots are their visible manifestations. At the beginning of the 11-year cycle, they appear at latitudes around 40°. As the cycle advances, spots form nearer and nearer to the equator. When plotted over time and latitude, the spots form a butterfly-wing shape whose area reflects the amount of activity in the cycle. When the 'wings' are at their greatest extent, at solar maximum, rotation is slightly modified, and the Sun radiates a little more energy.

A quadripolar model of the Sun is needed to explain epochs when solar activity is extremely low, confined to one hemisphere and only near the equator (the Maunder minimum). This configuration gives the poles the same magnetic polarity, while the equator is of opposite polarity. Solar cycles are then characterised by energy exchanges between three components: kinetic, magnetic and thermal. It should be understood that the transition from a dipolar to a quadripolar mode will affect the magnetic component, and, as a consequence, the thermal component. During normal cycles, changes in brightness, the index of the thermal component, are small, of the order of 0.1%. By contrast, during an anomaly the decrease in emitted energy has been estimated to be as much as 0.4%, and can thus explain the 'observed' impact on climate. This result is comparable to climatic models which predict that a doubling of greenhouse gases in the atmosphere would bring about an increase of 4 W m^{-2} (or + 1.5%) in energy received at the Earth's surface.

We should approach these simplified models and values with caution, as reality is far more complex. They do, however, show that the Sun is a variable star, to whose whims the Earth is very sensitive.

SPACE WEATHER

Interfaces between the corona and the solar wind, and between the interstellar medium and the solar wind, give rise to many questions of general interest to astrophysicists. These involve, in the first instance, the question of mass loss in stars, and, in the second, the question of the evolution of the interstellar medium. The region concerned is known nowadays as the heliosphere. Since the Earth and its magnetosphere are contained within it, it is vital in the space age that we study the nature of the heliosphere, and more specifically the variability of this extended solar atmosphere, as it bathes the whole solar system.

Let us now look at the origins and evolution of the science of space weather. At the end of the nineteenth century, Norwegian physicist Olag K. Birkeland definitively demonstrated that polar aurorae are the result of ionised particles from the Sun interacting with the Earth's magnetic field. In the 1930s, an English geophysicist, Sydney Chapman, put forward the hypothesis that magnetic storms affecting radio transmissions are due to particles, erupted from the Sun, arriving in the Earth's atmosphere. It might have been possible well before this to infer the existence of a continuous outpouring of particles from the Sun by studying the orientation of cometary tails, which point away from the Sun as a result of the 'pressure' of the solar wind. Only in the late 1950s did simplified models, such as those of the American Eugene N. Parker, show that the solar wind is connected with the corona and its very high temperatures.

It is obvious that pressure in the corona is greater than that in the interstellar medium; thus, in the absence of any resistance, solar material will escape into space.

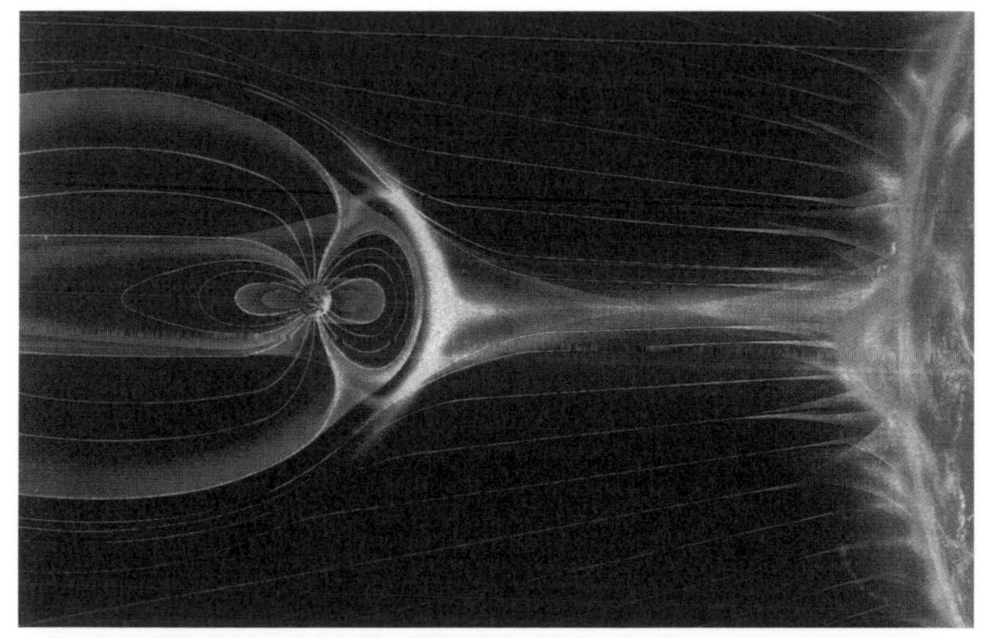

Fig. 1.14. The interaction of the solar wind with the Earth's outermost atmospheric layer, the magnetosphere. In the sunward direction, the solar wind compresses the force lines of the Earth's magnetic field. On the side away from the Sun, the magnetosphere exhibits a long 'magnetotail' extending over several thousands of Earth radii. (NASA.)

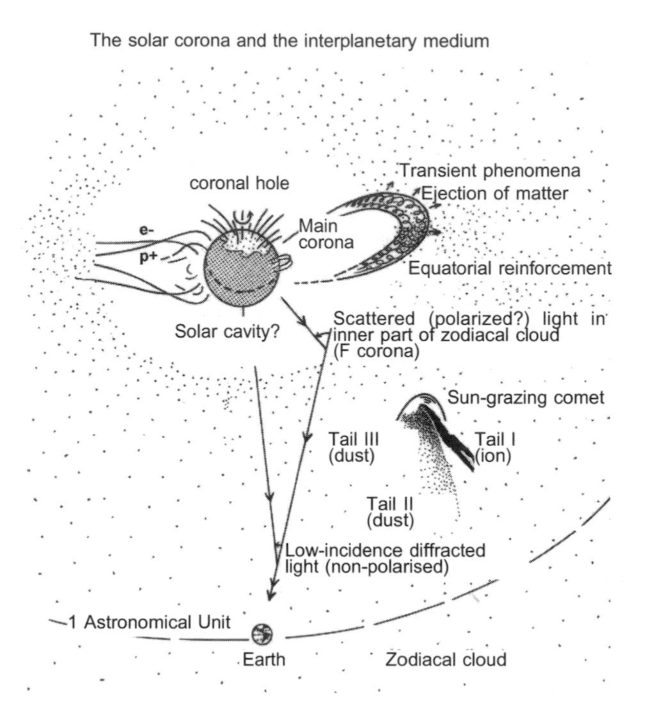

Fig. 1.15. Summary diagram of the Sun's environment, showing the various components of the interplanetary medium. (S.K.)

Near the Sun, the velocity of expansion is low, as there is a certain counter-pressure from the outer layers; further out, pressure from the upper layers falls off and the flux is increasingly accelerated by dynamic pressure.

Much is now known about the solar wind. Its velocity in the vicinity of the Earth is about 400–700 km s^{-1}, and its temperature varies between 50,000 and 500,000 K. Its density is only a few tens of hydrogen atoms per cubic centimetre. These data lead to the conclusion that a million tonnes of hydrogen escapes from the Sun every second. Over its 10 billion-year lifetime, the Sun will have lost in this fashion some 2×10^{18} tonnes of hydrogen, although this is only a negligible one-billionth of its mass. The energy expended in accelerating the solar wind is also negligible, representing only a millionth of the total energy expenditure of the Sun.

However, the Sun does not emit particles at an homogeneous density from all points on its surface. It is evident that emission is strongest in regions where the lines of magnetic force are open to outer space; that is, at the coronal holes. These regions, where temperatures are lower than in the surrounding corona, therefore lose less energy through convection. They appear darker, as a result of emitting less radiation in the form of light. Now, if the energy source maintaining the corona is more or less the same everywhere, it is to the kinetic energy of solar wind particles that we must look in order to investigate the particular energy budget of these regions.

On 21 August 1996, the SOHO satellite detected with its Lasco instrumentation a massive solar ejection event. In just a few moments, several million tonnes of gas and ionised particles (electrons and protons) were launched from the corona through a rent in the magnetic field, and were lost to space. This plasma cloud pierced the Earth's magnetic shield three or four days later, and the shock compressed the magnetosphere on the sunward side and dragged out a long tail on the opposite side. Particles infiltrated the magnetic field over the polar regions. In a zone known as the auroral oval, between latitudes 60° and 75° in either hemisphere, and at altitudes of between 100 and 1000 km, high-energy solar protons and electrons from the eruption interacted with the Earth's upper atmosphere to switch on one of Nature's most beautiful sights: the polar aurorae. Nitrogen and oxygen in the atmosphere, absorbing electrical energy, emit luminous photons of characteristic colours: red and green for oxygen, and blue and violet for nitrogen.

This auroral interaction may well have its effects upon humankind: for example, on 11 January 1997, the American telecommunications satellite Telestar 401, launched by AT&T, ceased transmitting as a result of a solar flare. About 50 satellites have now been damaged by such storms from space. Strong electric currents engendered by flares can overload major transformers and high-tension power lines, causing massive blackouts. In 1978 a large part of New York State was electrically paralysed for several hours. Similarly, on 13 March 1989, six million inhabitants of Quebec found themselves without electricity for ten hours. Today, astrophysicists are armed with a few satellites and ground-based telescopes which keep a constant watch on the Sun, and on its storms from the moment they are born to their arrival in the vicinity of Earth. They would like to know more about their origin, to make more precise predictions.

THE SUN AND SPACE MISSIONS

In the 1970s and 1980s, ambitious projects to observe the Sun were carried out at all wavelengths, some as part of manned missions like the Apollo Telescope Mount on the now defunct Skylab, the Salyut space stations and the Shuttle-borne Spacelab 2, and some on automatic satellites such as Helios, Solar-A, ISSE 3, OSO-8 and the Solar Maximum Mission (Solar Max).

Further developments of solar physics in space during the 1990s have included Ulysses, the European Space Agency's 'solar polar' probe, taken into space by the Shuttle; more limited instruments on a Spartan space platform used on Shuttle missions; Yohkoh (Japanese Institute of Space and Astronautical Science); Solar B; most notably, the ESA/NASA probe SOHO; and in the near future, the solar probe mission. NASA's Transition Region and Coronal Explorer (TRACE) is now aloft, and experiments will be mounted on the forthcoming International Space Station, whose survival may well depend on the observations!

Let us now describe the space programmes most concerned with the observation of the Sun.

Fig. 1.16. This view of Skylab in Earth orbit shows the module with its solar panels and, in the white area, the covers of the various observing instruments.

Skylab

Skylab was an experimental space station put into orbit by the Americans, to investigate the effects of prolonged stays in space, and the benefits which might be derived from work in the space environment. NASA had been planning an Earth-orbiting space station since 1965, but lack of funding led to the shelving of the project. Then, after the cancellation of three of the Apollo lunar exploration missions, it finally became possible to put together a space station, using material available from the Apollo programme.

The name Skylab was given to the project in 1970. It was planned to fly three teams of three astronauts each in the module. Apollo vehicles were used to ferry the astronauts to and from Skylab. Indeed, the main section of Skylab was a converted third-stage S-IVB section of a Saturn V launch vehicle. The interior of this module became a capacious living space and workshop. Other sections were added later, and space-walks were possible by means of an airlock module. Apollo vehicles could join with Skylab via a multiple docking adaptor.

In 1973, Skylab had on board telescopes and spectrographs for observations in the ultraviolet, extreme ultraviolet and even Hα, as well as an occulting coronagraph. More than 170,000 images of our local star were acquired with these instruments, and X-ray photographs were obtained from space. The mass of information gathered provided a great boost to solar physics, and decisive advances were made in aspects of coronal studies such as coronal holes, transition regions, and large mass ejections, all later consolidated during a whole series of international symposia. The 'Skylab era' will long be considered seminal in the history of solar physics.

Solar Max Mission

The Solar Maximum Mission satellite (Solar Max, or SMM) was launched on 14 February 1980, to keep watch on the Sun during a time of maximum spot activity. Only eight months into its mission, one of its instruments and its attitude control system failed. In April 1984, NASA mounted an unprecedented orbital recovery and repair mission to restore operational capability to SMM, although the time of maximum solar activity was long past. Shuttle astronauts, using the MMU (Manned Manoeuvring Unit), a kind of 'flying armchair', made perilous sorties into space to capture the failed satellite and replace defective equipment. Challenger Shuttle mission STS-41C began on 6 April 1984. On the third day into the mission, astronaut Dr George Nelson made his way with the MMU to the slowly rotating satellite, and tried to anchor himself to it with a specially designed grapple. The intention was for him to use his own MMU thrusters to counter the rotation of the satellite, which would then be taken on board the Shuttle by means of the robotic manipulation arm mounted on the cargo bay. But Nelson could neither lock on to nor slow SMM, and in fact set it tumbling even faster. Mission Control at the Goddard Space Flight Center finally managed to stabilise the satellite so that the manipulation arm could take hold of it and deliver it to the cargo bay, where, the day after, Nelson and his fellow astronaut James van Hoften worked to replace the defective parts.

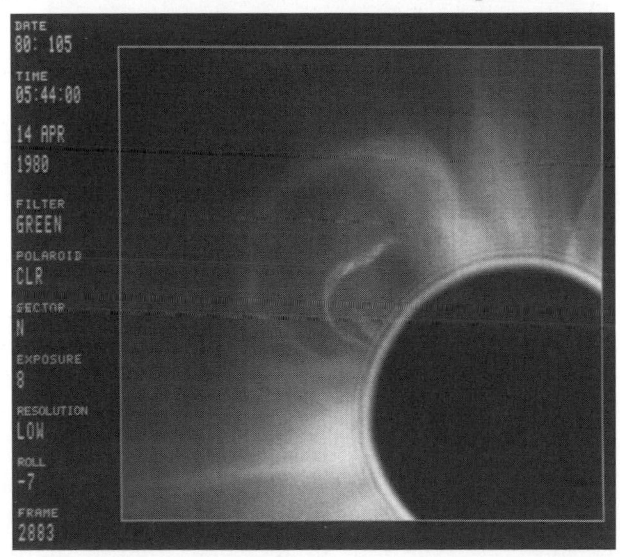

Fig. 1.17. A large mass ejection unfurling from the Sun, as seen in white light (green filter) with the externally occulting coronagraph of Solar Max in April 1980, near the time of peak solar activity. The larger loop is reminiscent of the leading edge of a spherical wave, inside which twisting elements of a prominence climb to great heights. The outer edge of the black occulting disk is at about 1.8 solar radii from the Sun's centre. (HAO and NASA.)

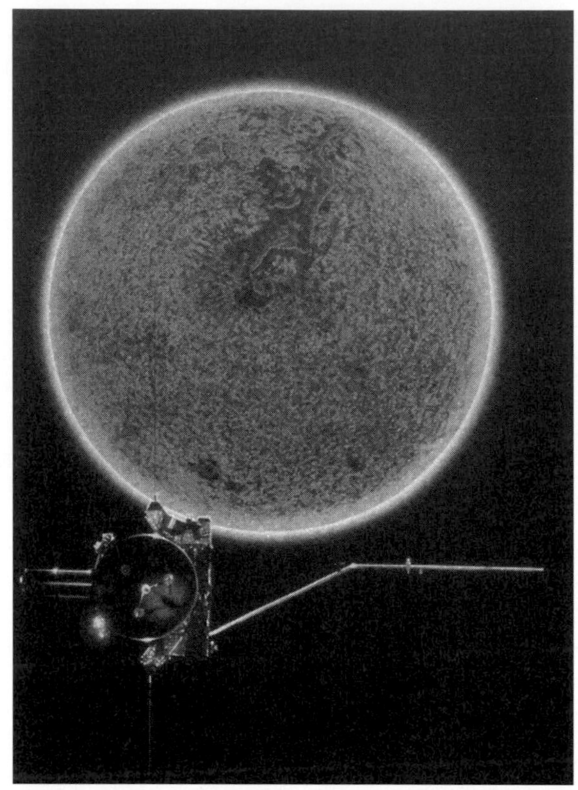

Fig. 1.18. Artist's impression of the Ulysses probe approaching the Sun. (ESA.)

Overcoming the problems of weightlessness and cumbersome gloves, they performed the repairs in a remarkably short time, and Solar Max was withdrawn from the bay and relaunched. Thereafter it functioned perfectly, and was soon transmitting excellent data. An increase in solar activity in the late 1980s caused a considerable rise in temperature, and expansion of the Earth's atmosphere at Solar Max's orbital height, increasing drag on the satellite and finally causing it to re-enter and burn up on 22 December 1989 – a victim of the solar activity it was designed to monitor.

Ulysses

The odyssey of this particular Ulysses, launched in 1990 by the Shuttle Discovery (Mission 36, STS-41), was to take it first to Jupiter, an encounter which would deflect it out of the plane of the ecliptic to overfly the Sun's south pole during May–November 1994, and its north pole during June–October 1995. One of Ulysses' objectives was to study the behaviour of the solar wind at the poles by *in situ* measurements of the components of the magnetic field, density, temperature, and so on, at various latitudes and radial distances.

The solar wind is a proton–ion–electron plasma which originates in the corona before being ejected into interplanetary space in both low-velocity (300–400 km s^{-1}) mode and high-velocity (up to 1,000 km s^{-1}) mode. Other research aims involved the study of the part played by magnetohydrodynamic (MHD) shock waves in causing discontinuities in the solar wind; the effects of solar acoustic waves on energy and plasma transfer phenomena in different layers of the Sun's atmosphere; and the creation of a global model of structures orchestrating the ever-changing activity of the heliosphere. In order to leave the ecliptic plane and fulfil its mission, Ulysses made use of a technique suggested in 1937 by Soviet researcher A. Sternfeld: a gravitational 'lever' or 'slingshot' for an indirect approach to the Sun. This involves flight away from our local star, towards Jupiter, which will then become a gravitational 'sling'. Precisely controlled manoeuvres by Ulysses took it on an incurving trajectory past the giant planet, a little of whose energy was utilised by the spacecraft to escape from the ecliptic plane.

Scientists were surprised to learn that the structure of the Sun's polar magnetic field was unlike that of the classic solar-wind models, and that, contrary to expectations, the field was much more heterogeneous than predicted. Ulysses' instruments revealed the existence of waves – probably a new type of MHD phenomenon – with periods of 10–20 hours, apparently of very slow variability. Could this be the key to the transfer mechanism of the magnetic field to space by the solar wind? Or was it the result of convective motions deep down within the Sun? The specialists are still debating the origin of the phenomenon.

The Ulysses mission has led to the classification of the solar wind into its two components: the high-speed stream issuing from coronal holes, and the more typical slower stream of the heliosphere's equatorial 'streamer belt'. The origin of coronal mass ejections (CME) remains a mystery.

Yohkoh

Yohkoh is the Japanese word for sunbeam. This satellite, from the Japanese Institute of Space and Astronautical Science (ISAS), was designed for high-energy observations of the Sun, and especially its flares and other coronal disturbances.

Carrying Japanese, American and British instruments, Yohkoh was launched in August 1991 on a three-year mission, later extended by another five years. Of fairly modest size (2 × 1 × 1 m and weighing 400 kg), the satellite follows a circular orbit at a height of 600 km, with a period of 97 minutes.

The payload consists essentially of four instruments: the SXT (Soft X-ray Telescope); the HXT (Hard X-ray Telescope); a Bragg Crystal Spectrometer (BCS); and a Wide-band Spectrometer. The SXT instrument – a joint product of the Lockheed Palo Alto Research Laboratory in the USA and the University of Tokyo – observes in the 4–60 Ångström waveband, with a pixel resolution of 1,700 km on the solar surface. It can image the whole Sun, or part of it, with a CCD camera for high-speed time sequences. The HXT investigates flares at wavelengths of 0.8–0.1 Ångströms. The BCS shows the complex lines of highly ionised elements, such as FeXXVI, FeXXV, CaXIX and SXX, using germanium-crystal detectors. This instrument examines the state of ionisation of gases during flares. Yohkoh's raw data are transmitted to Earth, and sent by the various receivers to the Kagoshina station of the Japanese Institute of Space and Astronautical Science, Tokyo, where they are analysed. Yohkoh's admirable images are made available on CD-ROM to research centres throughout the world, and can be found on the Internet. They have led to a greater understanding of the active corona, with its ionised streamers of gas forming, twisting and leaping into space along magnetic field lines.

Between 1991 and 1996, the SXT became the most prolific instrument ever used in solar physics, having taken over three million images. The first years of the Yohkoh mission were extremely productive, and many international symposia were devoted to discussion of its results and publishing of its discoveries about coronal eruptions and the dynamics of coronal structures. Thanks to this instrument, one of the authors was able to discover X-ray 'flashes' in polar holes, with associated microstreamers.

SOHO

The SOHO (SOlar and Heliospheric Observatory) project saw close liaison between ESA and NASA, as part of ESA's Cornerstone 1 Solar–Terrestrial Science Program. The SOHO platform was constructed in Europe by an industrial consortium led by Matra-Marconi, and instruments were provided by both European and American research laboratories. NASA was responsible for launch and operational facilities, and the antennae of their worldwide Deep Space Network (DSN) were used to guide the satellite into space. Mission Control was at the Goddard Space Flight Center, in Maryland. The 1.6-tonne observatory was launched on 2 December 1995 by an Atlas IIAS rocket from the Kennedy Space Center in Florida.

SOHO was first conceived as an observatory for the study of the Sun's internal structure, as well as its extensive outer atmosphere and the origins of the solar wind, whose ionised particles ceaselessly bathe the whole solar system. SOHO lived up to the best expectations of ESA and NASA, and gave unprecedently clear insights into

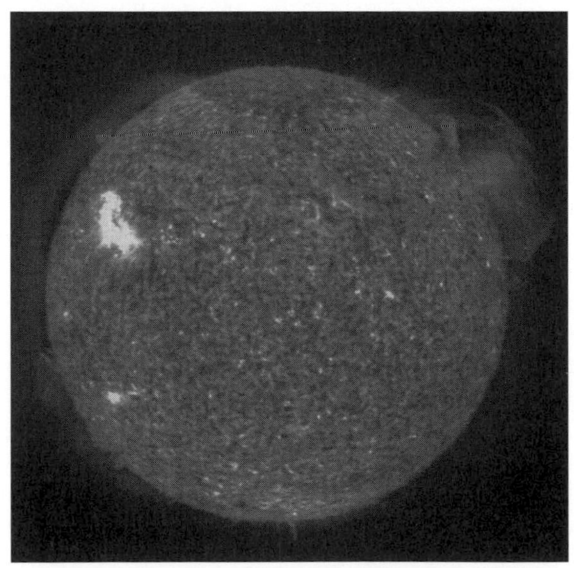

Fig. 1.19. A typical view of the quiet Sun in the light of HeII, obtained with the EIT instrument on board SOHO, 18 September 1997. (ESA/NASA.)

Sun–Earth interactions. Some of the Sun's best-kept secrets were revealed to scientists, giving clues to the heating of the corona, and the origins of the Sun's magnetism and the solar wind.

For the first time, researchers were able to study the Sun 24 hours a day, and 365 days a year. SOHO's uninterrupted view was made possible by a 'halo' orbit around a Lagrangian point, L1, 1.5 million km toward the Sun from the Earth, a point where the gravitational attractions of the Sun and the Earth cancel each other out. Obviously, this point is much nearer to the Earth than to the Sun, because of the enormous difference between their masses: the mass of the Sun is 330,000 times that of Earth. This peculiar elliptical orbit, known as a halo orbit, was an ideal vantage point for continuously observing the Sun, far from the vagaries of the Earth's atmosphere and the effects of its rotation.

Observation of the outer corona from Earth is possible only during rare and short-lived total eclipses of the Sun, and further study has to rely upon Earth-orbiting X-ray spectrographs and EUV instruments on board satellites equipped with coronagraphs. Even these devices can provide only partial clues for the creation of models to explain certain coronal structures. The formation, evolution and duration of such phenomena as coronal streamers, arches, loops or plasma flows, structured and driven by magnetic fields, seem to have their origins in permanent, small-scale dynamical phenomena and sudden larger-scale instabilities. To properly understand the physics of such structures, high spatial resolution is required over long timescales, justifying the use of a large, well-equipped satellite such as SOHO.

The same necessity for continuous study applies with the physics of the Sun's internal structure, and of acoustic pressure waves, discovered in 1960, and a whole new discipline – helioseismology, or solar seismology – has arisen. As has already been mentioned, helioseismologists study modifications in spectral lines resulting from ascending and descending motions of acoustic waves originating in the convective zone (200,000 km below the surface). Certain modes of oscillation, known as gravity modes, occur at frequencies impossible to measure from Earth. It is, therefore, easy to see why SOHO was programmed to operate principally in the realms of coronal physics and helioseismology. These two disciplines shared most of SOHO's payload, with five coronal physics experiments (Sumer, CDS, EIT, UVCS and Lasco, together with the Swan solar wind experiment) and three helioseismology experiments (Golf, Virgo and SOI-MDI). France made notable contributions, with Sumer, Golf, EIT, Lasco and Swan.

A great leap forward in solar physics was made with SOHO, and the results of the programme continue to pay dividends.

Cluster

Cluster was the second part of ESA's Cornerstone 1 'Horizon 2000' programme. Its aim was to study the behaviour of the Earth's magnetic field lines in three dimensions. Four satellites were involved, and they were to be placed in pyramidal formation within the magnetosphere over two years. This tetrahedral configuration would enable the satellites to move as much as 18,000 km apart, or to close to within 200 km of each other, in key areas of near-Earth space. The Cluster flotilla would move in highly elongated elliptical orbits with a perigee distance of 25,513 km and an apogee distance of 140,318 km. All four satellites were identical in size, mass (1,200 kg) and instrumentation (magnetometers, electron detectors and various other probes).

Data from Cluster would have furthered our knowledge of Earth–Sun relationships by studying the first interface encountered by the products of varying solar activity, where the particles of the solar wind buffet the terrestrial magnetic field. The four satellites, produced under the direction of the German Dornier organisation, were to be the first 'passengers' aboard the new European Ariane V rocket on its maiden flight (Flight 501). On 4 June 1996, at 14.36 Paris time, a trajectory fault entailed the destruction of the rocket and its payload. This launch failure, 61 seconds into the flight, temporarily halted the programme. A commission of enquiry, set up shortly after the incident, came to the following conclusions: 'Total loss of guidance and altitude information at 37 seconds after the beginning of the ignition sequence of the main motor (30 seconds after launch) caused the loss of Ariane 501. This loss of information was a result of specification and conception errors in the inertial reference system software ...' This short extract is taken from the joint ESA/CNES (Centre National d'Etudes Spatiales) communiqué of 23 July 1996. While not implicating the hardware of the positional control system, the commission's report made various recommendations – which were incorporated into the continuing Ariane programme – about the functioning of the launch vehicle in response to software. ESA has decided to launch further Cluster satellites.

TRACE

TRACE (Transition Region and Coronal Explorer) was launched on a Pegasus launch vehicle from Vandenberg Air Force Base in April 1998. The launch was scheduled to allow joint observations with SOHO during the rising phase of the solar cycle to sunspot maximum. TRACE explores the magnetic field in the solar atmosphere by studying the three-dimensional field structure, its temporal evolution in response to photospheric flows, the time-dependent coronal fine structure, and the coronal and transition region thermal topology. With these data, solar specialists expect to dispel the mysteries concerning coronal heating and impulsive MHD phenomena, with benefits not only to solar physics, but also to studies ranging from stellar activity to the MHD of accretion disks. The magnetograms produced by MDI on SOHO provide a complete record of the eruption and distribution of photosphere magnetic fields, which will be invaluable for understanding TRACE observations of coronal hole formation and coronal mass ejections. Both of these phenomena have profound effects on our space environment and the Earth's magnetic field.

THE FUTURE AND DEATH OF THE SUN

The Sun was born some 4.7 billion years ago, when, in one of our Galaxy's spiral arms, a cloud of interstellar matter, swept by shock waves from exploding stars, collapsed in upon itself under the influence of gravity. As the centre of the cloud contracted and rotated, gases – principally hydrogen – became condensed enough to trigger thermonuclear reactions, converting hydrogen into helium and creating the Sun of today.

3.5 billion years from now, the Sun will have consumed almost all of the hydrogen in its core. Its central regions will contract with increasing rapidity, as the failing thermonuclear energy processes allow gravity to overcome their outward thermal pressure. Pressure and temperature will rise as a result of this contraction, and the Sun's outer layers will swell enormously as this energy is evacuated. The photosphere will turn red. On the Hertzsprung–Russell diagram, the Sun will leave the main sequence and move into the red giant branch. During its passage along this branch, thermal pulsations will take place, and the Sun's diameter will increase to as much as one astronomical unit (1 AU = the mean distance from the Earth to the Sun = 150 million km). The solar wind will intensify, and our star will lose 40% of its mass. Its brightness will increase to as much as 5,200 times the present value, and Mercury will be engulfed. But Venus and the Earth will not share this fate, as the mass of the Sun will have fallen to 60% of its initial value, and the two planets will have migrated away to 1.2 and 1.7 AU respectively as the Sun's gravitational grip is relaxed. In the heart of our star, as temperatures soar to 100 million degrees, the fusion of helium atoms to produce carbon and oxygen will begin. This 'helium flash' will occur at the centre of the Sun. Hydrogen burning, producing more helium, will continue in a layer around the core. 7.7 billion years in the future, the Sun's central supplies of helium will be exhausted, and nuclear reactions will cease. As a result, the Sun will contract again. Its density will increase, triggering two types of nuclear reaction in two layers: at the periphery, hydrogen will again be converted to helium, and in a shell around the core, helium will be transmuted. This outpouring of energy will again cause the Sun to dilate, as thermal and gravitational pressures seek equilibrium. Eruptive phenomena, in the form of thermal pulsations, will occur. The intense radiation pressure will eject gusts of matter. The Sun will become a dwarf, with a radius only 3% of its present value. Its surface temperature will rise from 4,000 K to 100,000 K, and its colour will pass from red to white. In this final white-dwarf stage, the Sun, now no bigger than the Earth, will illuminate the gases it has ejected, shining on for a few billion more years, before growing dark.

The mass of matter ejected into the cosmos, some of which will consist of atoms from the Earth, will be carried off by solar winds, some day to become part of new stars, new planets, and possibly other living creatures.

2

How to observe the Sun

To admire and photograph the Sun's surface, all that is needed is a small telescope, a few accessories and some elementary precautions. Prominences, spots and faculae are all remarkable phenomena, and can be of great beauty. What is more, the assiduous amateur observer can further our knowledge of the Sun by collaborating with professionals working in ground-based facilities or with space platforms.

OBSERVING THE PHOTOSPHERE IN WHITE LIGHT

A modest telescope will put the amateur within reach of phenomena of the photosphere such as spots and their penumbrae, and granulation. However, there are certain vital precautions to be taken. The amount of energy emitted by the Sun, and consequently falling upon the Earth's surface, is such that 99.99% of the radiation must be eliminated before safe observing is possible. To set against this, the Sun's very brightness, and the huge size of some of its phenomena, mean that observation with even modest instruments can reveal much fine detail on its surface.

Taking into account price, reduction of the effects of turbulence, and resolution, the optimum diameter for a refracting telescope for amateur observations is from 80 mm to 150 mm (approximately 3–6 inches).

Precautions for solar observing

As important as the choice of instrument is the acquisition of effective eye protection. Any direct observation without such protection will result in certain damage to the eye.

The Sun's radiation can have several harmful physiological effects upon the eye. The dangers of ultraviolet radiation, especially in mountainous regions, are well known. Infrared rays are equally dangerous, since they can cause coagulation of the retina as a result of heating.

Ophthalmologist Dr G. Quintel has carried out a study of retinal damage brought about by careless observation of the Sun. He states that, when observing a luminous object, the diaphragm of the iris can contract to provide partial protection to the eye,

Fig. 2.1. Alice and François observe the partial eclipse of 12 October 1996 in safety through welder's glass. This filters out ultraviolet and infrared, and vastly reduces the Sun's glare. (V. Guillermier.)

to a minimum diameter of about 1.5 mm in most people. If the light is not too strong, dazzling without permanent lesion will occur, and the eye recovers after a while. With light that is too intense, however, heating effects will overcome the eye's capacity to cool itself by means of blood circulation through the retina, and the increase in temperature within the tissues will bring about permanent lesion. However, no sensation of pain is felt, as the retina possesses no nerves to transmit it. Damage to the unprotected eye will be proportional to the amount of luminous flux experienced, and the duration of the exposure. With inadequate eye protection, repeated observation of the Sun kills cells at the centre of the retina, and in the long term will lessen visual acuity to a considerable degree. It is therefore absolutely necessary to have effective protection, even during the briefest of observations, or if the slightest discomfort due to dazzling is felt. Paradoxically, being dazzled may give the impression of comfort, since the visual response is diminished. The most basic precaution is to ensure that any filter used is not only effective for visual light, but also filters out ultraviolet and infrared radiations.

We draw the reader's attention to the dangers of the kind of filter – usually red or black and inscribed with the word 'SUN' – which often comes as an accessory with an astronomical telescope. These filters screw into the front end of 24.5-mm (~1-inch) eyepieces, which means that they are close to the focal point and may become very hot. They may shatter after a few tens of seconds, exposing the observer to the concentrated and therefore dangerous light of the Sun.

There are three more or less convenient methods for observing the photosphere.

The projection method

This is the cheapest and most commonly employed method. The image of the Sun's disk is projected onto a white screen placed between 10 and 30 cm behind the eyepiece, and the brightness of the Sun is such that sunspots and faculae will be readily seen. Under no circumstances must the observer look through the eyepiece, but only at the projected image, which will not be so bright as to be harmful.

To point the telescope at the Sun without using the finder – which should already have been covered to ensure that its internal reticule, near the focus, is not overheated – all that needs to be done is to look at the shadow of the instrument on the ground. When the shadow is at its smallest, the Sun's rays must be shining directly down the tube, and the Sun's image will be centred for projection. In order to avoid heating lenses in the eyepiece by moving them too close to the focal point while focusing, it is advisable first to focus upon some distant object on the horizon before aiming the telescope at the Sun. The projection method will reveal dark sunspots with their surrounding greyish penumbrae, and, by contrast, faculae glowing brightly on the photosphere. Limb darkening, which lends the image a pleasing sense of depth, may also be seen. This phenomenon is caused by the absorption of light from regions near the Sun's surface by cooler and therefore less luminous gases situated above the photosphere.

There are several major drawbacks with the projection method. Observers are advised not to use Schmidt–Cassegrain telescopes, since heating of the optics may damage the adhesion of the secondary mirror to the corrector plate. Also, photographing the Sun as a projected image is difficult because the telescope prevents the camera from being held in the optical axis, and photographs of the image taken from the side will be distorted.

The Herschel wedge

The Herschel wedge is a right-angled accessory tube containing a block of glass with non-parallel faces, or a prism, which allows 95% of the light to pass through, with the rest reflected towards the eyepiece. The eyepiece should be fitted with a dark filter of observational quality. There are some drawbacks with this method. The Herschel wedge has an opening through which most of the light and heat passes, and the observer must beware of being burned. Also, as with the previous method, all the solar flux enters the instrument, causing considerable heating. This is why the accessory should not be used with Schmidt–Cassegrain telescopes (see above). It is also difficult to use it with Newtonian reflectors, as the focal point in these instruments is close to the eyepiece tube, and focusing may be impossible if the device is present. The main advantage of the Herschel wedge is that it is transferable from one instrument to another.

Full-aperture filters

The best method, if not the cheapest, is to use an optical-quality neutral filter over the objective lens of the telescope. In this position, heating will not be a problem, and safe observations can be carried out for long periods. The full-aperture filter consists of a sheet of glass, with scrupulously flat surfaces which do not distort the light rays entering the instrument. A metallic coating (aluminium or Inconel) on the glass will filter out all but 1/100,000 of the light. Prices will, of course, be higher for the larger filters. Some observers use full-aperture filters made of mylar, although this produces somewhat degraded images, and colours and definition usually suffer. Moreover, it is possible for some ultraviolet radiation to find its way through these thin barriers, with possible damage to the eye. Mylar should therefore be used only for photographic work.

Welder's glass No. 14 is cheap and easy to obtain from builders' merchants and specialised suppliers, and allows observation of the Sun in complete safety.

THE WOLF NUMBER (RELATIVE SUNSPOT NUMBER)

A record of the Wolf or Zürich Number is systematically kept by the Royal Observatory of Belgium in Brussels.

A simple method of counting the number of sunspots is to project the Sun's image onto a white paper screen upon which is drawn a circle representing the Sun. The circle may be marked with heliographic coordinates. By convention, north is at the top and west to the left, and it is therefore necessary to orientate the projected image of the Sun, with the equator in the correct place. The screen is turned until a spot is seen to run parallel to the pre-drawn equator. Then, recording of spot positions may begin. This operation is not as easy as might be assumed: instability of the screen support, wind effects, and the very act of marking the spots' positions, cause unwanted movement and lack of precision. Precautions to take include

- Minimising the duration of the marking operation to two, or at most, three minutes.

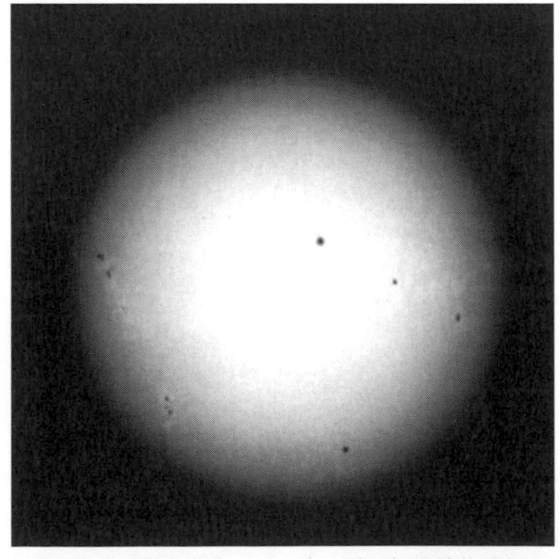

Fig. 2.2. Photograph of the Sun taken at the Pic-du-Midi Observatory with a 10-cm finder telescope, of focal length 1.5 m, with a violet filter. (S.Koutchmy).

- Using a soft lead pencil or fine fibre pen.
- Finalising the sketch, by drawing in the outlines of the spots and their penumbrae, away from the screen support.
- Increasing the stability of the instrument's mount.
- Using a motorised equatorial mounting, since a solar image 120 mm across moves at 1 mm per second.

Heliographic discs in current use include those of diameter 139 mm (on the equator, 1 mm on the disk corresponds to 10 000 km on the Sun's surface), and those of diameter 114 mm (on the equator, 1 mm corresponds to 1 heliographic degree).

It is necessary to orient the Sun's projected image using solar coordinates. As seen from Earth, the Sun's polar axis may appear to be leaning to the left or to the right, and the north and south poles to be inclined towards or away from us according to the time of year. This is because the Sun's equator is inclined at 7° 11′ 30″ from the plane of the ecliptic (the plane of the Earth's orbit), and so the Sun's orientation appears to change throughout the year. Around 6 March the south pole is at its maximum inclination towards Earth, and around 8 September the north pole is presented to us to the maximum extent. The inclination P of the solar axis in relation to the geographical north–south direction, and the variation of heliographic latitude of the centre of the solar disk, known as B_0, is given in the *Handbook* of the British Astronomical Association and in other publications. By allowing a spot to drift across the screen, the geographical east–west axis can be determined.

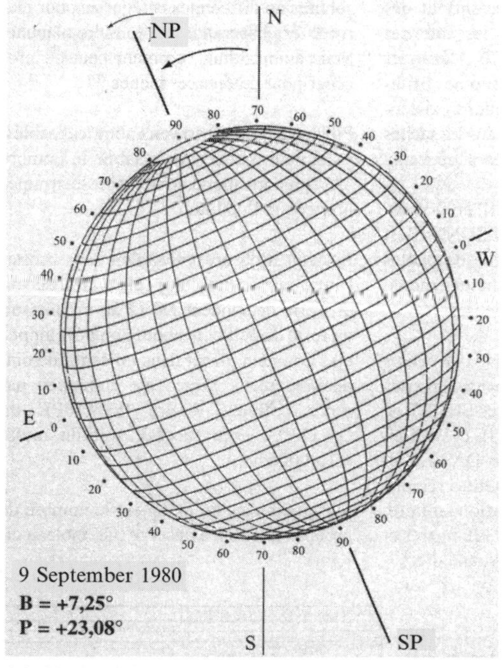

Fig. 2.3. Template of heliographic coordinates for the solar projection screen (*Eclipse* magazine, No.5).

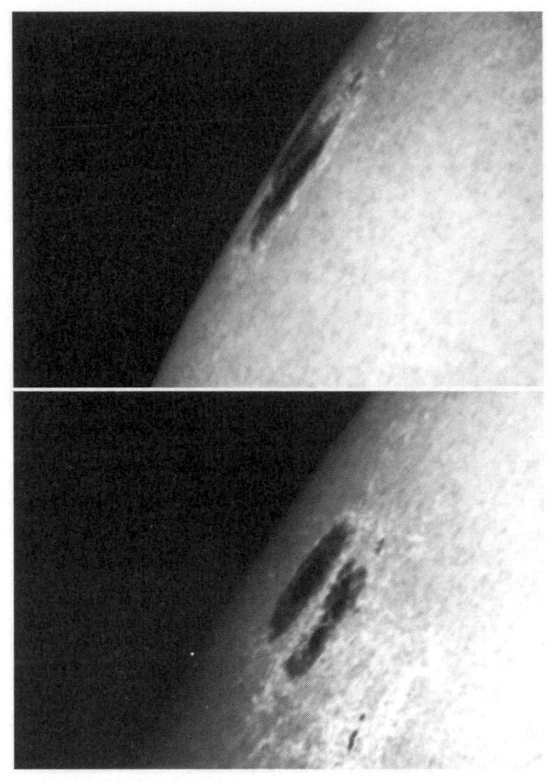

Fig. 2.4. Two photographs of the same spot group 24 hours apart, showing its apparent displacement across the solar surface. (SPO/NSO.)

PARAMETERS OF PHOTOSPHERIC ACTIVITY

As well as sunspot numbers, there are two other parameters which provide an insight into solar activity.

Total sunspot surface calculation

This value may be affected by errors, as erroneous estimates of positions and apparent surface area of spots near the limb are likely. The total sunspot surface is expressed in millionths of the visible solar hemisphere.

Proper motion of sunspots

Sunspots migrate in both latitude and longitude across the solar surface. Longitudinal displacement is usually more in evidence than latitudinal displacement. A sunspot's apparent motion results from both the Sun's differential rotation at the latitude of the sunspot, and the sunspot's proper motion. This latter tends to move leader sunspots (which are further west) away from their followers (further east). In the first case, apparent motion adds to the differential rotation, while in the second it subtracts from it. The study of the proper motion of sunspots is of importance in our understanding of mechanisms of the Sun's activity which are not yet completely explored. The very mass of the data in the possession of amateurs, compensating for inaccuracies in plotting, means that they have an important contribution to make.

OBSERVING PROMINENCES

In 1930 the French astronomer Bernard Lyot invented an ingeniously designed device: the optical instrument known as the coronagraph. This allowed observation of one of the Sun's most impressive phenomena – the prominences. A few years later, in association with J. Leclerc he made a film entitled *Flames of the Sun*, showing gigantic eruptive prominences complete with explosions, ejections and condensations of matter, as recorded with the coronagraph. Amateur astronomers intending to observe the Sun in similar fashion should not expect this to be the normal state of affairs, as most prominences are quiescent, and therefore evolve only slowly. In fact, Lyot and his colleagues used carefully selected sequences, and accelerated the most spectacular events when editing their film. It is, however, in the area of rapid phenomena that amateurs can carry out useful work, by recording or photographing extremely short-lived events (such as eruptive prominences) which may escape professional observers. Events worth recording may be of two kinds:

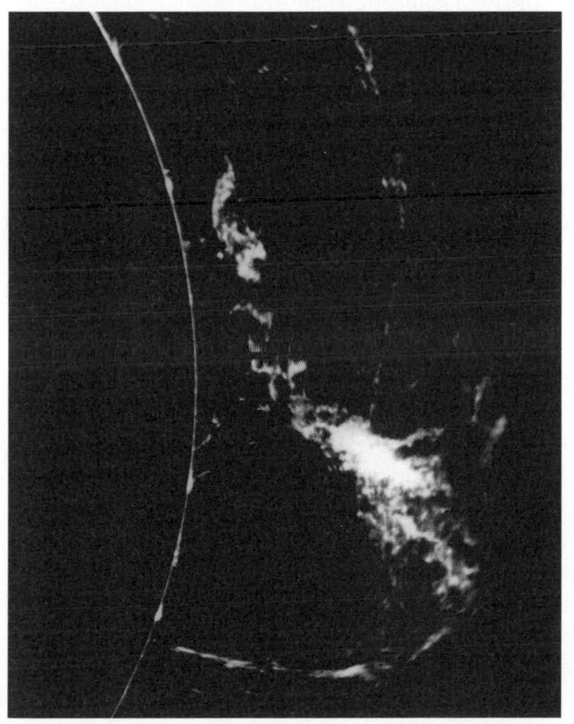

Fig. 2.5. A fine prominence observed in Hα light, using a coronagraph. (SPO/NSO.)

- Rapidly evolving prominences with ejecta ('spray' or 'surge' events). Here, evolution may last from about one hour to about ten hours. In active regions, during periods of maximum activity and flares, loops and structures with bright material will appear. With luck, very spectacular interlaced loops will be observed a few hours later, as a flare takes place at the Sun's limb.
- Sudden disappearances of prominences. Quiescent prominences, which are often at altitude, have a tendency to disappear once per rotation over very short periods of about half an hour, and may reappear few days later. However, if the local magnetic field has been disturbed by the onset of a new active region, they may not reappear.

If you wish to start observing prominences, available methods include the coronagraph, the Hα filter, and the spectroheliograph.

The Lyot coronagraph

The main function of a coronagraph is to completely block out light coming from the photosphere. Invented by the French astronomer Bernard Lyot, this delicate instrument uses a single lens fitted with a circular occulting disk at the focal plane, together with a cleverly conceived apodisation system situated beyond the occulting disc. The image of the solar disk is removed, whilst all around, the image of the corona is superimposed upon the diffraction ring and the sky background. A lens placed behind the internal occulting disk forms the image of the pupil before the Lyot diaphragm, and the spectral filter and further optics form the image on the detector. The coronagraph is a fairly complex instrument, and quite delicate both in optical fabrication and when in use. Microscopic dust particles on the optics, or a trace of mist in the Earth's atmosphere, can scatter the light to such an extent that the image of the corona will be badly spoilt. Coronagraphs are, therefore, often used at mountain sites where the air is purer. Moreover, the objective of a standard coronagraph consists of only one lens, which must be polished with great care. The light from the blue sky is still obstructive even at high altitude, and coronagraphs can be used to study only the inner corona and cool Hα emissions, which may observed out to some distance from the Sun. Coronagraphs incorporating highly polished silicon mirrors are now being developed in France, the USA and Russia.

A small Lyot-type coronagraph, produced by the Baader company, has been available to amateurs for some ten years now. Unfortunately, there are certain drawbacks with this accessory: it can be costly, and its use is limited to telescopes of certain focal lengths and with precision equatorial mounts, motorised on both axes. The coronagraph consists of a metal cylinder, attached to the 36.4-mm threaded focusing tube. It has four components, the main one being the internal occulting disk, which consists of a steel cone with a multi-layered coating of aluminium. The cone is placed at the focal point of the main objective lens, where the image of the Sun is formed. Its purpose is to occult the solar disk, simulating an eclipse. In order to cover the angular diameter of the Sun's disk, which changes throughout the year, the manufacturer offers six cones of different diameters. The occulted image of the Sun then passes through a field lens, which produces an image of the objective at a

Fig. 2.6. In the foreground is the 26-cm Lyot coronagraph of the Pic-du-Midi Observatory. This instrument, equipped with a photoelectric polarimeter, was used for ten years by J.-L. Leroy for measuring magnetic fields in prominences by means of the Hanle effect. J.-C. Noëns and collaborators are currently observing the corona in Hα with the same instrument. (Pic-du-Midi Observatory/OMP.)

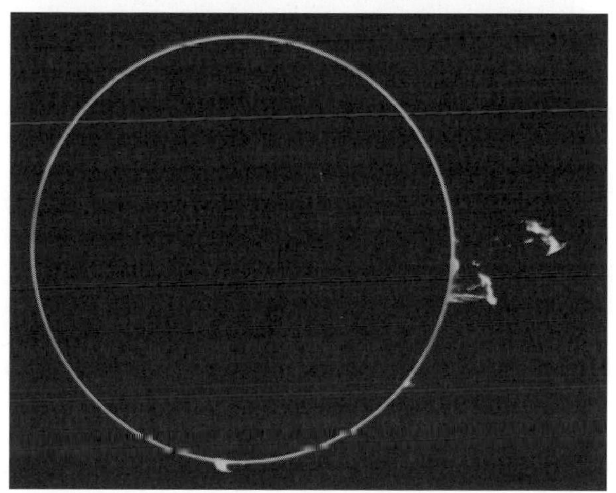

Fig. 2.7. Coronagraph image, 5 October 1988, 21h 35m UT. The coronagraph uses a highly polished 4-cm mirror as its objective. A prominence is shown in Hα. (Serge Koutchmy at SPO/NSO.)

diaphragm of variable aperture. The final element is the interference Hα filter. This filter transmits the emission line of ionised hydrogen, in which prominences are seen, and blocks out light from the photosphere. The more selective the filter (the narrower its bandpass), the better the image obtained. Baader offers two filters: the first (bandpass 1 nm) produces good-quality images; the second (0.4 nm) is more selective and produces images with greater contrast, though is obviously more expensive. Beyond the filter, the second objective forms the image and multiplies the

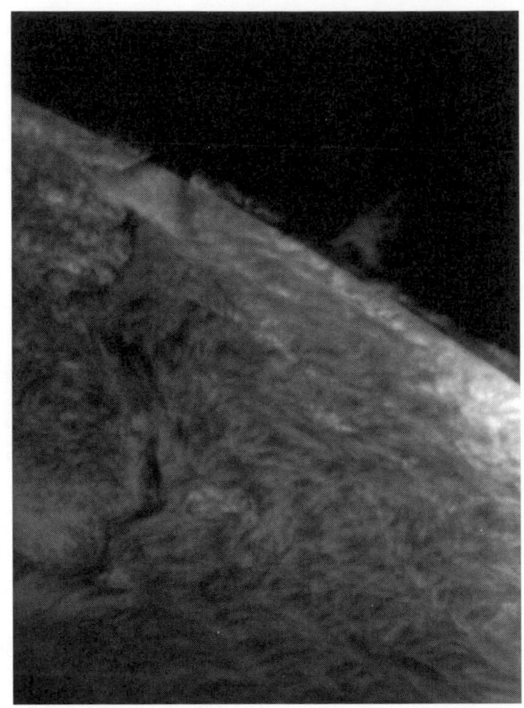

Fig. 2.8. The Sun's limb seen in Hα, in the vicinity of a filament which is seen as a prominence beyond the disc. Note also the chromospheric fringe. (SPO/NSO.)

initial focal length of the telescope by 1.5. With long-focal-length instruments, the solar disk is seen in its entirety; with shorter focal lengths, it is necessary to use an off-centring device to observe the Sun's limb.

The Hα filter

Another possibility is to use a prefilter and an interference Hα filter in series. An interference Hα filter transmits light at a wavelength of 656.28 nm (hydrogen emission line) with a bandpass of at least 0.07 nm. When used in conjunction with a prefilter, it allows observation of the chromospheric network and its magnetic structures, sunspots, faculae, flares, prominences and filaments, across the whole solar disc. The narrower the bandpass, the better the contrast. The position of the bandpass on the spectrum will depend on the temperature of the filter, and with narrow-bandpass filters in thermostatically controlled housings, this is maintained to within 1°. It is necessary to carry out adjustments to explore the profile of the line in question, and this is achieved by inclining the filter within its housing by means of an adjustment wheel. Shifting the wavelength slightly in either direction reveals solar structures of high radial velocity, by virtue of the Doppler shift. The prefilter must be used to eliminate ultraviolet and infrared emissions in order to ensure prolonged and safe observing. (Narrow-bandpass 'T-Scanner' Hα filters are marketed by the American firm DayStar.)

'PRO–AM' COLLABORATION IN SOLAR OBSERVATION

As we have seen, serious observing of the Sun and its many phenomena (flares, ejections, prominences, filaments, the corona, and magnetic fields) require not only a great deal of equipment, but also a measure of perseverance and systematic analysis.

Professionals have long known that observations are best carried out in the context of extended programmes of collaboration between observatories, on an international scale. More modestly, amateurs may collaborate with professionals at certain observatories (for example, Pic-du-Midi) where coronagraphs are used by selected amateurs. Observing programmes may involve small teams of amateurs, who undertake exciting observations using the most modern equipment, and contribute to the solar database, to the benefit of the scientific community.

In the future, amateurs will doubtless have access to space instruments such as those aboard SOHO, as part of a joint observing programme with collaboration in analysis and interpretation of results.

(If you are interested in observing the Sun, contact: in the UK, the Solar Section of the British Astronomical Association (BAA), whose reports appear in the Association's *Journal*; in France, the *Commission Soleil* of the Société Astronomique de France (SAF) which publishes the magazine *L'Astronomie*; and in the USA, the Sunspotters Club of the American Association of Amateur Astronomers (AAAA). Details appear in the addresses section at the end of this book. You will be put in touch with well-informed amateurs and professionals who can help you with your work. Your observations can be useful to the whole scientific community.)

OTHER SOLAR PHENOMENA

The Sun sends its radiation to Earth at all wavelengths. The effects of its light in our atmosphere create various phenomena of great beauty. For example, scattering of the Sun's light by molecules in the Earth's atmosphere causes the sky to appear blue (Rayleigh scattering). Ionisation and recombination of nitrogen and oxygen molecules produce the polar aurorae. The following are some phenomena originating from interactions with solar radiation.

The zodiacal light

The zodiacal light is caused by reflection of the Sun's light on interplanetary dust spread along the ecliptic plane. This phenomenon takes the appearance of a cone of faint light stretching above the horizon at night, and should not be confused with the Milky Way. It is visible in the west an hour after dusk, and in the east before dawn, especially in spring and autumn when the angle between the ecliptic and the horizon is at its greatest as seen from mid-latitudes. The shape of the cone shows the inequality of distribution of the interplanetary dust. Visually, the zodiacal light may be 25° across at its base, and the apex of the cone may be 60° above the horizon. From dark sites near the equator – where the ecliptic always stands well above the horizon – this light may invade the whole sky as the zodiacal band. Interplanetary

Fig. 2.9. Photograph of the zodiacal light from Sacramento Peak, at 2,800 m, using a simple camera of 35-mm focal length, at f/4, on an equatorial mount. Exposure time, 5 minutes. (K. Jockers and S. Koutchmy.)

dust grains do not remain in fixed positions, as solar radiation pressure pushes the smallest particles outwards, and gravity attracts the largest towards the Sun. This dust is constantly replenished by debris from comets orbiting within the solar system.

Photography of the zodiacal light is simple, with large apertures (f/2.8), and lenses of focal length 28–75 mm, in conjunction with a fast film (400–1000 ISO). Exposure times may be 2–10 minutes.

The gegenschein

The gegenschein (German, 'counterglow') is a very faint glow in the night sky, on the ecliptic opposite the Sun. It has a similar origin to the zodiacal light. Here, reflection of the Sun's light takes the form of an ellipse about 20–30° across. This phenomenon is extremely elusive, being best seen at latitudes around 40–50°, towards midnight from February to April and from September to November.

The rainbow

Rainbows are colourful daytime atmospheric phenomena, and are familiar to us all. They appear when the Sun's light shines into raindrops, and are caused by the difference between the refractive indices of air and water. Refraction involves a change in direction of light-rays as they pass from one medium to another. White light penetrates the raindrops, and is refracted. The angle of refraction depends upon the wavelength of the radiation, and so the Sun's light will be split into its different component colours. Within the raindrop, light is reflected at precise angles at its spherical inner surface, and then refracted again towards the exterior. If the light experiences only one reflection in the drop, a primary rainbow will be created. In rarer cases where there are two reflections, a secondary rainbow – fainter and with a reversed sequence of colours due to the second reflection – will be seen outside the primary bow. The mean angle of the primary bow around the Sun–observer line is 42°, and that of the secondary is 52°.

The green flash

At sunrise or sunset, when the atmosphere is particularly transparent and if certain other conditions apply, the upper edge of the Sun will fleetingly appear as a green 'flash' at the time of complete occultation by the horizon. The origin of this phenomenon lies in the fact that strongly refracted solar rays have to traverse a dense layer of the atmosphere at sunrise and sunset, before arriving at the observer, account being taken of the curvature of the Earth. The atmosphere, through which the light is passing tangentially, acts as a prism, bending some rays of light more than others – atmospheric refraction. In order of curvature, from greatest to least, are seen violet rays, then blue, green, yellow, orange, and red. Moreover, the atmosphere scatters colours of shorter wavelength (blue and violet) in all directions, while transmitting long wavelengths (orange and red). Herein lies the reason for the familiar image of a reddish Sun at sunset or sunrise. Refraction bends red and yellow rays only slightly, and they will be the first to disappear at the horizon. Conversely, blue and violet rays are scattered, and are therefore unseen by the observer. Only the green ray is sufficiently curved, and unscattered by the atmosphere, to be able to reach the observer as the Sun's upper edge disappears.

Parhelia (sundogs)

Parhelia (Greek *para*, 'beside'; and *helios*, 'Sun') are two patches of visible light at the same altitude as the Sun, and at 22° on either side of it. A parhelion is caused when the Sun's light is reflected by ice crystals in high-altitude clouds, such as cirrus.

Solar halos

Solar and lunar halos appear as luminous circles around the Sun and the Moon. They occur when all the droplets or ice crystals in the cloud masking the Sun or the Moon are of almost equal dimensions. Halos are composed of concentric rings, from blue-green on the inside to reddish on the outside. The diameter of these aureoles is about four times that of the Sun or the Moon. If all the droplets are of similar dimensions, the colours will be more distinct. The greater the difference in their diameter, the more the wavelengths will be spread and the more the colours will be indistinct. White lunar halos have a radius of about 20°.

Polar aurorae

As we have already mentioned, aurorae originate through interactions between the corpuscular radiation of the Sun (the flux of particles flooding into space from the Sun) and the Earth's magnetic field. Ionisation of oxygen and nitrogen, and their recombination with emission of red and green light, are the direct cause of aurorae. They take many forms. The most striking displays resemble vast, constantly shifting curtains of light, revealing the Earth's magnetic force lines. The phenomenon may last for only a few minutes, or all night.

Aurorae are seen mostly in high latitudes, near the poles. They can be seen from the more northerly areas of the UK and Canada, and occasionally from the south. If the faint glow of an aurora is seen, it is worth watching, as it may be a prelude to a fine display, with characteristic greenish tints, and reddish, slowly undulating curtains.

To photograph an aurora, set up a camera on a stable tripod, and take exposures of 5–20 seconds' duration. A 28-mm or 50-mm lens is best, with the widest possible aperture. Films of 400–1000 ISO are suitable.

EQUIPMENT FOR OBSERVING

Refractors, reflectors, coelostats, spectroheliographs: what are the characteristics of each instrument?

Tripods, altazimuth mounts and equatorial mounts

It is vital that any observing instrument – be it a pair of binoculars, a refractor or a reflector – is securely mounted. A pair of binoculars or a terrestrial telescope may be firmly held on the type of tripod used for photography, with various movable joints allowing it to be inclined in any direction.

An astronomical refractor, or a reflecting telescope, should rest upon a mount with two axes of movement. The altazimuth stand or mount has one axis vertical and the other horizontal (as with the largest solar telescope (THEMIS) on Tenerife).

If the horizontal axis of an altazimuth mount is inclined to match the angle of the observing site's latitude, parallel to the axis of rotation of the Earth (with the axis pointing at the Pole Star), an equatorial mount, or head, results. The polar axis is graduated in hours and minutes of right ascension. The other axis is the declination axis and is graduated in degrees. The advantage of the equatorial mount is that by adding a small motor to the polar axis, turning at a rate of once every 23 h 56 m in the direction opposite to the Earth's rotation, the motion of the stars in relation to the turning Earth is effectively cancelled out. To compensate for the rotation of the Earth *vis-à-vis* the Sun, the motor would need to turn once every 24 hours. If an altazimuth mount is used, the Earth's rotation will cause a star to drift out of the field of view, and it will be necessary to operate both axes to keep it central in the field. With equatorials, however, the motor will drive the polar axis and the star will remain in the field. The Moon has its own motion which does not keep time with the diurnal rotation of the sky, and to follow it accurately the declination axis may also be motorised to compensate.

The refractor

If you wish to observe sunspots in some detail, then a refractor (so called because of the refraction or bending of light within it), will perform the task. The refractor has a lens (the objective) which focuses rays of light to a point known as the focus. A smaller lens (the eyepiece) is used to examine the image formed by the objective.

The magnification of an astronomical telescope is given by dividing the focal length of the objective lens by that of the eyepiece. A small refractor of 50 or 60 mm diameter will allow magnifications of x50–100. Beginners naturally want to magnify as much as possible, but a telescope will not comfortably magnify more than 2.5 times its diameter in millimetres; so a 60-mm refractor will magnify well up to x 150. The usual magnification would be 1–2 times the diameter in millimetres (x60–120 for the 60-mm refractor).

Many lenses have faults, known as aberrations, of which spherical aberration and chromatic aberration are examples. Spherical aberration may be corrected by using aspherical lenses. Chromatic aberration is caused by rays of different wavelengths, or different colours, being focused to different distances, and to counter it, achromatic, apochromatic and fluorite objectives are used.

The advantages of the small refractor include its portability, the quality of its optical surfaces, its enclosed tube which cuts down turbulence, and the absence of any obstruction at the centre of the tube, which might cause diffraction. Images are sharper than with equivalent reflectors, and refractors give the best performance for the quality of the atmosphere at a given observing site. The performance is determined largely by the diameter and quality of the objective, and its focal length.

The biggest refractor at present used for solar work is in the domed tower at the Pic-du-Midi Observatory. The instrument has a diameter of 50 cm, and is equipped with an excellent spectrograph.

The reflector

The reflector uses a concave mirror to bring light rays to the focal point, and there are various types. The oldest design, dating from the early 1660s, seems to be that of James Gregory, a Scottish mathematician, but the construction of such an instrument was beyond the technology of the day, and it was not until the late 1660s that Isaac Newton simplified Gregory's concept and built the first reflector.

The Newtonian reflector has a parabolic main (or primary) mirror which collects, reflects and brings together light rays to a focus. A flat secondary mirror, mounted at 45° on the optical axis, reflects the light out of the optical axis of the primary mirror towards the observer.

Many optical combinations have been tried in search of the best performance from a reflector. The Schmidt–Cassegrain uses a main spherical mirror with an additional optical element – the corrector plate – at the front of the telescope. This glass plate corrects spherical aberration of the mirror. A secondary, hyperbolic mirror, usually attached to the plate, reflects the light rays back towards the focus. This system means that telescopes can be compact, and easily transported. There are also other configurations such as the Cassegrain Dall–Kirkham, the Ritchey–Chrétien combination and (the most popular) the Maksutov combination. It is difficult to choose between a reflector and a refractor. Reflectors are usually cheaper than refractors of equivalent diameter.

The spectroheliograph

Invented independently by the American George Ellery Hale and the Frenchman Henri Deslandres in 1891, the spectroheliograph records images of the entire solar disk on plate or film, at a chosen wavelength. It is primarily a spectrograph – an instrument designed to isolate a narrow band of the field of observation through a slit, and to disperse this unidimensional element as a function of wavelength by means of a diffraction grating. Moreover, the spectroheliograph has an entrance slit and a selective exit slit. By simultaneously moving the Sun's image across the entrance slit and the film placed behind the exit slit, an image of the entire solar disk can be built up at the chosen wavelength (for example, the Hα line at 656.3 nm, or the calcium K line at 393.4 nm). The spectroheliograph is not limited to the observation of filaments and prominences; the evolution of spots and faculae may also be studied. Of most interest, however, is the ability of this instrument to provide photospheric magnetograms via analysis of circular polarisation in a line split by the Zeeman effect, produced by a magnetic field.

The coelostat and the heliostat

Unlike astronomical telescopes – which are multipurpose instruments sometimes adapted for solar observation by the addition of neutral-density filters – a Lyot coronagraph or a Hα filter, the coelostat is specifically designed for studying the Sun, involving the benefits of long focal length and wide dispersion.

Coelostats have the ability to remain fixed not only on the image of the point on the line of sight, but also on images of all the points in the field, so there is no rotation of the image during observation. A plane mirror is mounted parallel to the Earth's polar axis, and rotates about this axis at half the diurnal rate. The image

Fig. 2.10. Some of the experiments set up by the Paris-Meudon Observatory at Nejapa-Oaxaca (Mexico) for the total eclipse of 7 March 1970. Behind Z. Mouradian can be seen the 50-cm flat mirror of his coelostat. (S. Koutchmy.)

everywhere in the sky is fixed. Incoming light is reflected to a secondary mirror, positioned according to the declination of the observed field but remaining fixed as diurnal motion proceeds. Such coelostats are widely used in solar instruments, in conjunction with large spectrographs. Many such instruments are housed in solar towers (such as those at Mount Wilson, Potsdam, Rome, Arcetri near Florence, and Meudon), where the coelostat is situated tens of metres above the ground to escape turbulence. The best-known example of a polar siderostat is at Kitt Peak, Arizona, where an image of the Sun is viewed at the base of the tower, with one or more spectrographs in a shaft below, sheltered from variations in temperature. The famous solar tower at Sacramento Peak in New Mexico, at an altitude of 2,800 m, has a different arrangement: 42 m above the ground at the top of the tower is an azimuthally mounted heliostat with two mirrors. The objective is a mirror 1.6 m across (76 cm useful aperture), placed at the bottom of a 60-m shaft. To obtain a fixed image of a large area of the Sun, the whole telescope/spectrograph tube is turned about a vertical axis, the 230-tonne assembly being suspended on a floating ring in a bath of mercury. The whole space through which the light beam passes is completely evacuated of air to minimise turbulence. The resulting images are of excellent quality.

Fig. 2.11. Sacramento Peak Observatory, New Mexico, at an altitude of 2,800 m. In the centre is the vacuum tower, and to the left, the dome housing the coronagraphs. Note the forest of fir trees, which reduces thermal exchanges and contributes to the high quality of images obtained. (SPO/NSO.)

THE OBSERVING SITE

A site from which to observe the Sun must be chosen with care, if the best images are to be obtained. A telescope set up on a surface such as concrete, flagstones, asphalt or gravel will produce degraded images of the Sun. Such surfaces create much turbulence because they readily re-radiate heat. On the other hand, a wooded area with grassy surfaces will contribute to good-quality images with little turbulence, as will a site surrounded by water or fresh snow.

The best times to observe are from one to three hours after sunrise in summer, when turbulence is still minimal. Conditions may also be favourable when the Sun is at its highest in the sky, since, at this time, the layer of air through which its rays pass is at its thinnest.

Turbulence within the observing instrument may be lessened by allowing into it only the amount of light necessary for the observation. Atmospheric turbulence may be reduced by choice of a suitable site and observing time. However, in spite of all precautions, even in fair conditions resolution will not be greater than about 1 arcsec. This means that a 150-mm telescope, with its theoretical resolving power of less than 1 arcsec, is more than adequate for systematic study of the Sun's activity. An increase in telescope diameter will not bring about any improvement in the image, but will entail the risk of degrading it because of turbulence due to thermal convection currents within the tube. If a coronagraph is used, the set-up is of even greater importance.

Observation of a region of the sky near the Sun while the Sun itself is masked behind, say, a tree or a chimney, will reveal the brilliant white atmospheric halo.

Fig. 2.12. Pic-du-Midi observatory in the 1960s. In the foreground is the Baillaud dome, where Bernard Lyot carried out his admirable observations. (S. Koutchmy.)

Dust particles, droplets and ice crystals suspended in the air cause this phenomenon. We see here the ability of large molecules to scatter solar light almost uniquely in the direction of the flux of incident light, known as specular scattering. To lessen the atmospheric halo effect, it is best to avoid sites where the air may be laden with dust (cities, plains) and to refrain from observing when certain atmospheric conditions prevail. This will include fine, hot days when masses of tropical air produce heat haze, or days when cirrus cloud in front of the Sun produces a small, very bright halo close to it. The best solution is to move to higher ground, as aerosols are concentrated in lower layers of the atmosphere, and their intensity diminishes by a factor of at least three for every 1000 m in altitude. To enjoy a good-quality sky, it is best to observe from a site well above the murky layers of the atmosphere, which are capped by the temperature inversion layer normally lying at a daytime altitude of 2.5 km.

SOLAR PHOTOGRAPHY

Observation of the solar surface can be satisfactorily undertaken with a 150-mm instrument, and the same is true for photography. Excellent high-resolution photographs of the Sun can be obtained with a 200-mm reflector or a 150-mm refractor. With this type of instrument, a full-aperture filter with an attenuation factor of density 4 should be used, corresponding to the transmission of only 1/10,000 of incident light. Even this is ten times more light than is transmitted by filters used for visual observation.

The luminosity of the Sun corresponds to a magnitude of –26.74 – 1.5 billion times brighter than the planet Venus. This vast brightness means that the slowest commercially available films can be used: these include, in black and white, Ilford Pan F at 50 ISO, Agfa 25 ISO, or Kodak TP 2415, which is more sensitive and produces good contrast. Colour slide films include Kodachrome 25 ISO and Fuji Velvia 50 ISO, with their slow emulsions, good resolving ability and wide contrast. For prints, Kodak Ektar 25 ISO, Konica 50 ISO or Agfa Ultra 50 ISO are recommended. Exposure times may be from 1/250 to 1/2,000 s, which reduces the effects of camera shake and turbulence, and obviates the need for motorised equatorial tracking.

The apparent diameter of the Sun varies between 31′ 32″ and 32′ 36″, with a mean diameter of 32′ 04″, or 9.32×10^{-3} radians. If we employ the optical formula relating the apparent diameter of a small object, α, at infinity, and its image D obtained with an optical system of resultant focal length f_{res}:

$$D = f_{res} \,.\, 0.0093$$

where D is the diameter of the Sun on the film in millimetres, and f_{res} is the resultant focal length of the instrument used, in millimetres, it can be simply calculated that the diameter of the Sun on the film will be 9.3 mm per metre of focal length.

Details of the solar surface such as 'rice grains' and sunspots can be photographed with focal lengths of at least 2,000 mm, to obtain an 18-mm image of the whole of

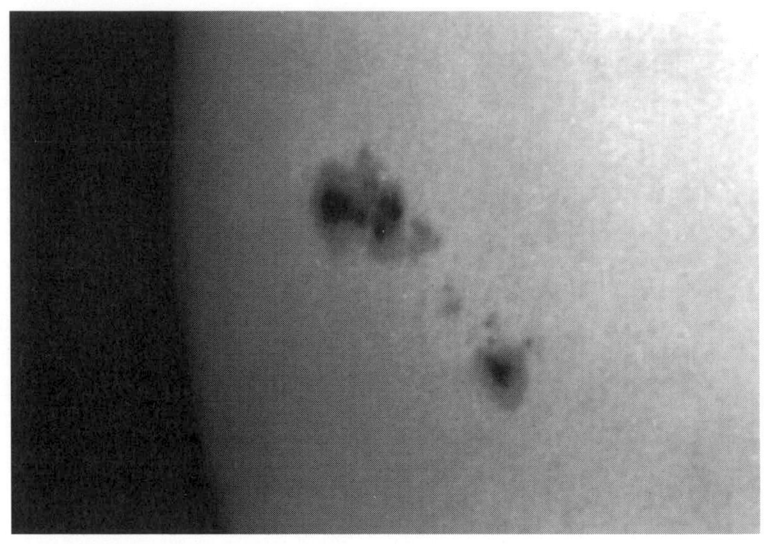

Fig. 2.13. All that was needed to photograph this sunspot group near the Sun's limb was a 60-mm refractor of focal length 700 mm, with eyepiece and SLR camera. Note the faculae, and the limb darkening which gives a striking three-dimensional effect. (M. Marrel).

the Sun, in 36 × 24-mm (35-mm) format. At focal lengths of over 2.5 m, only part of the Sun can be photographed, and close-ups of sunspots can be obtained. All such photographs will require a full-aperture filter – or a mylar screen, which will not produce such good results – to diminish the intensity of the Sun's light.

Prominences may be photographed with a Lyot coronagraph or a Hα filter. The camera is put in the place of the eyepiece, and a T2 ring adaptor is used to attach the camera to the telescope. Exposure times may be bracketed between 1/15 and 1/125 s, and films of 50–200 ISO should be used.

SOLAR SPECTROSCOPY

Spectroscopic analysis is a very broad subject, from which a great deal can be learned. As the Sun's light is so bright, it is possible to disperse it very widely and examine its spectrum, with the individual lines indicating the presence of different chemical elements. A diffraction grating (usually 10 cm long with at least 1,200 lines per mm), and collimators of at least 3 m focal length, are needed. Apertures are adapted for use with a particular telescope, which must be of long focal length for good resolution, through the entrance slit, of fine details of the solar surface, such as sunspots.

Doppler effects can be recorded. These include asymmetrical shifts in lines, or splitting of lines due to relative velocities of elements; or the Zeeman effect, with symmetrical shifts and polarisation effects involving electrons attached to nuclei,

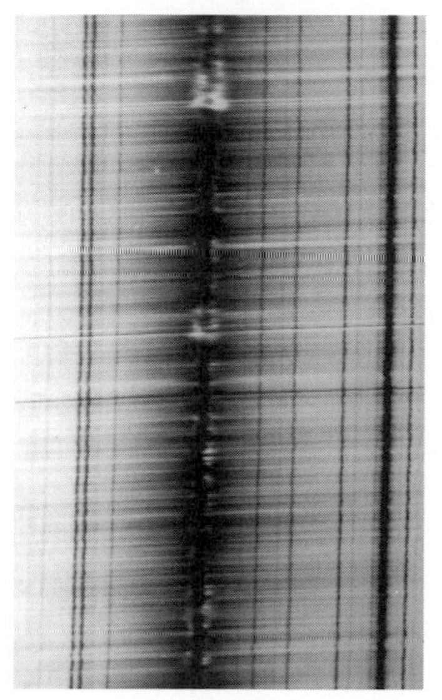

Fig. 2.14. Part of the spectrum around the K line of CaII, obtained near the centre of the Sun through a slit 0.5 arcsec wide and 250 arcsec long. Note the Doppler effects, due to turbulent convection, on the fine photospheric lines. (SPO/NSO.)

under the influence of magnetic fields, producing lines as a result of changes in state. Other phenomena within the reach of a good-quality spectrograph include the granulation resulting from turbulent convection, the Evershed effect (outward flow of material in the penumbrae of sunspots), and the differential rotation of the surface.

It is, however, rare for an amateur to assemble such a system, which is nevertheless to be found in all major solar observatories. The comparatively recent introduction of CCD cameras in spectral imaging has led to great progress, and computerised systems based on the properties of spectrographs are now being used for rapid cartography of velocities and the magnetic field. Such multidimensional analyses are now being carried out with increasingly sophisticated software.

3

Eclipses of the Sun and Moon

Throughout its history, the human race has tried to understand, and calculate with increasing accuracy, the motions of Sun, Moon and Earth. Ra, the Sun god, depicted with the head of a hawk upon which shone the solar disk, led the hierarchy of the Egyptian pantheon, and had charge of many agricultural and administrative events. From Johann Kepler and Isaac Newton onwards, the study of planetary motions entered an era of precise calculation and simulation. But how do the Sun, Moon and Earth move? What is the nature of an eclipse of the Sun, or of the Moon? What characterises the different stages of these eclipses? What different kinds of eclipse are there?

ECLIPSES AND CELESTIAL MECHANICS

The Moon moves around the Earth. The Earth rotates on its axis and moves around the Sun. Only a few centuries ago, these commonplaces were by no means evident. Great numbers of observations, and great minds, were required to put positional astronomy on a firm footing. In the sixteenth century, Nicolaus Copernicus, and in the seventeenth century, Kepler, Galileo and Newton, set the Sun in its rightful place at the centre of the system which bears its name. Nowadays the motions of the heavenly bodies are well understood, and with computers the fundamental work of mathematicians such as Laplace, Lagrange, Gauss and Poincaré forms the basis of extremely accurate calculations.

The motions of the Sun

The Sun contains 99.9% of the mass of the solar system, and therefore controls the movements of all its members. The Sun rotates on its axis, which is inclined at 82° 49′ from the plane of the ecliptic (the plane of the Earth's orbit) and its mean rotation period is 25.4 days. It also exhibits differential rotation: 24.6 days at the equator, and 35 days at the poles.

The star we call the Sun, with its family of planets, is travelling at 19.7 km s^{-1} towards the point known as the solar apex, situated between the stars ν and ο in the constellation of Hercules. The Sun also moves with its local group of stars at 250 km s^{-1} around the centre of the Galaxy.

The motions of the Earth

The motion of the Earth as it rotates on its polar axis causes the alternation of day and night. The apparent motion of the sky caused by the Earth's rotation, with rising and setting of the Sun and other bodies, is called diurnal motion. The time taken by the Earth to rotate on its axis, with reference to the stars, is known as a sidereal day (23 h 56 m 4.091 s).

The other motion of the Earth is its orbital journey around the Sun. This causes the apparent drift of the Sun across the background stars of the zodiac. This apparent path – the ecliptic – is the projection of the Earth's orbit upon the 'map' of the sky. The time taken by the Earth to orbit the Sun, with reference to the stars, is known as a sidereal year – 365 d 6 h 9 m 9.5 s. The time taken by the Earth to return to the same point in its orbit is called a tropical year – 365 d 5 h 48 m 45.975 s. The difference between the sidereal year and the tropical year arises from precession (described below).

The Earth's axis of rotation is not perpendicular to the plane of its orbit. The polar axis is inclined at 23° 26′ from the perpendicular to this plane. This inclination causes the Sun's rays to strike our planet in varying directions throughout the year, and results in the changing seasons and inequalities in the lengths of day and night.

The flattening of the Earth at the poles and its equatorial bulge cause irregularities in the attraction exerted by the Sun and the Moon. These two bodies act upon the polar axis, causing it to describe a cone of 23° 26′ around the perpendicular to the plane of the ecliptic over a period of 26,000 years, and to swing between angles of 21° 59′ and 24° 36′ over 41,000 years. This movement, known as the precession of the equinoxes, is overlaid and complicated by other irregularities and interferences such as nutation.

The motions of the Moon

The Moon's orbit is inclined to the plane of the ecliptic by 5° 8′ 43″, and therefore crosses it at two points, called nodes. So the Moon is sometimes above the ecliptic plane, and sometimes below it. When the Moon crosses the ecliptic travelling from the southern to the northern hemisphere, it passes the ascending node, and as it crosses southwards it passes the descending node. The axis between these nodes, called the line of the nodes, moves in a retrograde fashion during the Moon's revolution. The line of the nodes turns once in a period of 18 y 11 d – a period known as the saros.

The motion of the Moon through the sky causes it to be seen at varying angles in relation to the Sun. The illumination of the Moon from different angles brings about the familiar phenomenon of phases. When the night side of the Moon is turned towards us and is therefore invisible from Earth, the phase is called new Moon. At first quarter, we see half of the sunlit side. When the Moon appears completely illuminated and circular, we see the full Moon. Finally, at last quarter, we see the illuminated surface on the opposite side of the Moon to that seen at first quarter.

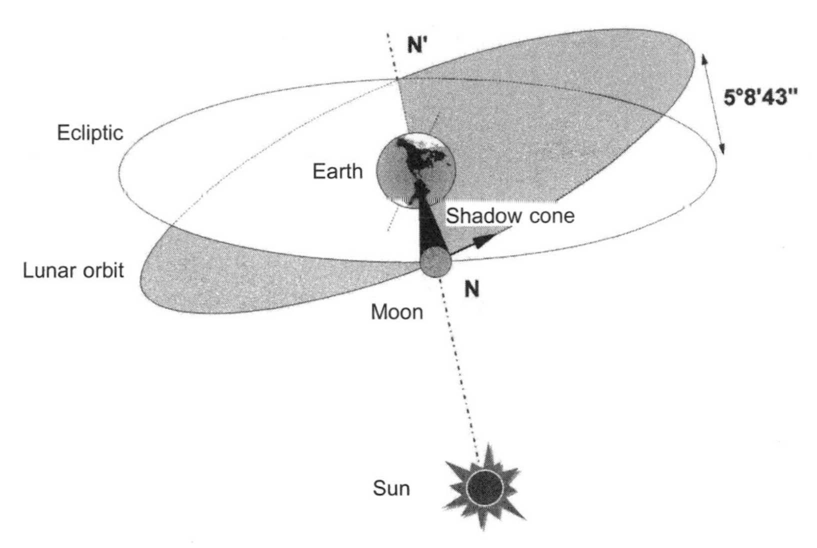

Fig. 3.1. At position N there is only a small difference between the geocentric longitudes of the Sun and the Moon. Moreover, the longitudes of the Moon and of the node (ascending, in this case) are similar. An eclipse of the Sun occurs. In position N′ the geocentric longitudes of the Sun and the Moon are 180° apart, and the longitude of the node (here, descending) is similar to that of the Moon. An eclipse of the Moon occurs. (P. Guillermier.)

Thereafter, the waning crescent signals the approach of another new Moon.

The cycle of lunation, or synodic period, lasts approximately 29 d 12 h 44 m 3 s. It should be noted that between two consecutive passages of the Moon through perigee there is a different interval called the anomalistic period, lasting 27 d 7 h 43 m 11 s. The difference between these two intervals is caused by the fact that the Earth is orbiting the Sun.

So, after the Moon has completed one anomalistic revolution, it has to move a little further to present the same phase as seen from Earth. In the course of one lunation the Moon therefore moves further than one revolution. These two intervals – orbital period and anomalistic period – have a common multiple which was known to the Chaldeans: the saros, lasting 18 y 11 d 8 h, provided that four of these years are leap years. If there are five leap years, then the saros lasts 18 y 10 d 8 h. This period is close to 223 lunations, 239 anomalistic revolutions and 242 passages of the Moon through the ascending node. As it orbits the Earth, the Moon rotates on its axis. The time taken by the Moon to rotate once is equal to its orbital period, which explains why the Moon always presents the same face towards the Earth. However, the Moon's axis of rotation is not strictly perpendicular to the plane of its orbit: it is inclined at an angle of between 83° 11′ and 83° 29′ from this plane. This means that at certain times we can see beyond the north pole, and at other times beyond the south pole. Otherwise expressed, the mean centre of the Earthward face appears to shift through ± 6° 44′ in a north–south direction. This phenomenon, which was

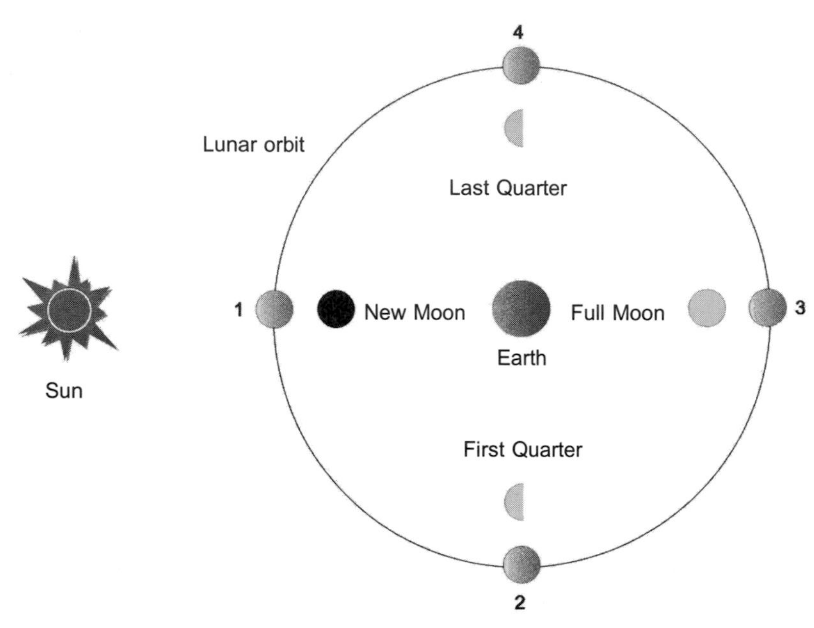

Fig. 3.2. Diagram explaining the phases of the Moon, and Earthshine. The Moon's appearance as seen from Earth is shown inside the circle representing its orbit. The diagram also helps us to understand the phenomenon of earthshine: it is seen near new Moon, and is due to light reflected from the daytime side of the Earth falling upon the night side of the Moon. (P. Guillermier.)

discovered by Galileo in 1636, is known as libration in latitude. Our satellite also accelerates and decelerates in its elliptical orbit, obeying Kepler's laws, and therefore does not move with constant velocity. This allows us to observe an extra strip of the Moon 7° 45′ wide beyond both east and west limbs successively – a phenomenon known as libration in longitude, discovered by Hevelius in 1654. A third libration, of about 1°, is caused by the rotation of the Earth, and is known as diurnal libration.

Observers will not see exactly the same area of the Moon at the beginning of the night and at the end of the night, as they have been carried around with the turning Earth. The sum of all these librations means that, from any point on Earth, it is possible to see 59% of the Moon's surface.

Celestial motions and eclipses

The word eclipse comes from the Greek *ekleipsis*, meaning abandonment or omission, echoing the fear of the ancients that their Sun god had forsaken them.

In their orbital dance, Earth, Moon and Sun may move into alignment. If the Earth lies between the Sun and the Moon, the Moon will pass into the Earth's shadow, and an eclipse of the Moon will be seen. If the Moon lies between the Earth and the Sun, the Moon is seen to move across the face of the Sun, and an eclipse of the Sun occurs. In strict terms, this is an occultation of the Sun, since the Sun is only masked by the Moon and continues to shine. On the other hand, an eclipse of the

Moon is a true eclipse, as the light normally falling upon it is blocked, and the Moon ceases to 'shine'.

We now examine in more detail the positions of the three bodies at the times of these special events.

Celestial mechanics and eclipses

On the subject of calculation in celestial mechanics, and the beauty of eclipses, S.A. Mitchell wrote, in his book *Eclipses of the Sun* (1923): 'Amongst all the wonders of all the wonderful sciences there is no science which deals with such a gorgeous spectacle as is exhibited by the queen of the sciences, astronomy, at the moment when the Earth is gradually shrouded in darkness and when around the smiling orb of day there appears the matchless crown of glory, the so-called corona. Nor can any science duplicate the wonderful precision shown by the work of the astronomer in his capacity to predict hundreds of years in advance the exact hour and minute at which an eclipse will take place and the locality on the Earth's surface where such an eclipse will be visible.'

Indeed, for 5,000 years astronomers have been turning their minds to the task of celestial prediction.

During the nineteenth century, canon Alexandre Pingré calculated the dates of all the eclipses of the previous 3,000 years. In 1887 the German astronomer Theodor von Oppolzer published his catalogue, entitled *Canon der Finsternisse*, with the elements of 8,000 eclipses of the Sun from 1208 BC to 2161 AD, and 5,200 eclipses of the Moon from 1207 BC to 2163 AD. In 1900, the mathematician Schram revised

Fig. 3.3. As Earth, Moon and Sun line up, the Sun's atmosphere is revealed. Image of the corona obtained from Siberia on 9 March 1997. (M. Molodensky and colleagues, Izmiran, Moscow.)

Oppolzer's latitude and longitude positions for the lines of centrality, and more recently, H. Mücke and J. Meeus, and Fred Espenak, have published fuller and more elaborate canons than Oppolzer's. By the third millennium BC, Babylonian astronomers had calculated that, after 223 lunations, the Moon returns to the same point in its orbit with reference to its perigee. This period – the saros – is therefore used to predict eclipses, since, once an eclipse is observed, there will be another after one saros cycle, as the Moon will have reached the same position.

Although the saros is not exact enough for the geographical circumstances of an eclipse to be calculated, it will certainly give the date when it will occur. However, since the saros period is eight hours more than 6,585 days, the eclipsed Sun will be seen in a region 120° further west than was the previous eclipse, after one saros. A complete set of eclipses needs to recur 70 times; that is, 70 saros periods or 13 centuries, to be repeated over the same area on Earth. In any one saros, the number of eclipses may be between 39 and 47 for the Sun, and 39 and 47 for the Moon. The cycle of variations lasts about 590 years, and the average number of eclipses in a saros is 43 for both Sun and Moon.

Having discussed frequency, let us now turn to the distances, diameters and positions of the Earth, the Sun and the Moon. The respective distances of the Sun from the Earth, and of the Moon from the Earth, are approximately 149,600,000 km and 384,000 km, the former being about 400 times greater than the latter. The diameter of the Sun is 1,391,000 km, and that of the Moon is 3,476 km – again a difference of about 400 times. This extraordinary coincidence means that the two bodies have nearly equal apparent diameters in the sky. (Table 3.1 shows the maximum, mean and minimum apparent diameters of the Sun and the Moon.) Thus, the Moon can pass in front of the Sun, and hide it for a short time. The Sun is eclipsed. This type of eclipse occurs when the Moon lies between the Sun and the Earth, at new Moon. If the apparent diameter of the Moon is equal to, or greater than, the apparent diameter of the Sun at this time, the Moon will hide the Sun and the eclipse will be total. If the Moon's apparent diameter is less than that of the Sun, the lunar disk will not succeed in hiding the Sun completely, and at the moment of maximum coverage of the Sun's disk by the Moon, a ring of sunlight will remain around the black lunar disk, producing an annular eclipse. If celestial mechanics dictate that the Sun's disk is only partially hidden, the event is a partial eclipse.

Table 3.1. Apparent diameters of the Sun and the Moon.

Apparent diameter	Maximum	Mean	Minimum
Sun	32′ 35″	32′ 01″	31′ 31″
Moon	33′ 31″	31′ 05″	29′ 22″

The phenomenon of the solar eclipse is easily explained. A cone of shadow (umbra and penumbra) stretches away from the Moon on the side opposite the Sun. The length of the umbral cone is between 325,000 km and 375,000 km. When the shadow is intercepted by part of the Earth, an eclipse of the Sun will be observed

Fig. 3.4. The Moon's shadow sweeps across the Earth's surface on 11 June 1983. Note the dark patch at the centre of the photograph. (NASA.)

from that part. The eclipse will be total for observers in the umbral shadow, and partial for those in the penumbra. The trajectories of the umbra and penumbra across the Earth's surface are curves several thousand kilometres long, and the diameter of the umbral shadow may be as much as 269 km. At the equator, the shadow moves at a speed of 1,706 km h^{-1} along a path across the globe, the medial line of this path being referred to as the centre line. The penumbra falls upon an area 3,000 km wide around the umbral shadow.

All the above factors combine to limit the maximum duration of totality of a solar eclipse to 7 m 58 s at the equator, and 6 m 30 s at latitude 45°. The likelihood of a total eclipse occurring for a given place on the Earth's surface is once every 370 years.

An eclipse of the Moon occurs when the Earth lies exactly (or almost exactly) between the Sun and the Moon, at the time of full Moon. If the Earth and the Moon both orbited in the same plane, there would be an eclipse every month at full Moon; but as we have seen, the Moon's orbit is tilted with reference to plane of the orbit of the Earth, and eclipses cannot occur with such frequency. Only when the Moon is near the nodes will eclipses take place, if the Sun is aligned with both Moon and Earth.

The length of the Earth's umbral cone varies with the changing distance from the Earth to the Sun. At perihelion it reaches out to 1,386,000 km – more than 108 times the Earth's diameter. At aphelion, its length is 1,433,000 km – more than 112 times the Earth's diameter. The diameter of the part of the Earth's shadow traversed by the Moon also varies with the Earth–Moon distance.

If the Earth is at its mean distance from the Sun of 1 AU (149.6 million km), then:

- if the Moon is at perigee (nearest to Earth, at a distance of 356,400 km) at the moment of the eclipse, the diameter of the Earth's umbral shadow equals 9,503 km, while the penumbra is 16,069 km across;
- if the Moon is at apogee (furthest from Earth, at a distance of 406,704 km), the diameter of the Earth's umbral shadow equals 9,056 km, while the penumbral diameter is 16,525 km;
- at the Moon's mean distance of 381,500 km, the Earth's umbral shadow is 9,244 km in diameter, and the penumbra is 16,334 km in diameter.

Obviously, the Moon will fit completely within the umbral and penumbral shadows, as its diameter is only 3,476 km. Moreover, the maximum duration of a lunar eclipse (1 h 45 m) can be predicted from the fact that the Moon moves at an average of 30 arcmin per hour, and its mean apparent diameter is 31 arcmin. A total eclipse of the Moon is therefore visible – unlike a solar eclipse – from a large part of the Earth's surface.

ECLIPSES OF THE SUN

During a partial or annular eclipse, the Sun takes on a very unusual aspect: the Moon hides part of its visible disk, and the familiar circular form is lost, while the heat we receive from it is diminished.

During a total eclipse, a succession of events takes place. The most spectacular of these is totality, when the Moon has hidden the Sun's disk, revealing the corona, which shines brightly. Even though it is a million times less bright than the disk, the corona is far brighter than the surrounding 'darkened' sky and can be as bright as a full Moon. Much can be learned from observation of these phenomena.

The significance of solar eclipses

Why do astronomers travel the world to see eclipses? Firstly, because the Sun may be likened to a physics laboratory in space. In its vicinity, conditions of temperature, pressure, gas density, and so on, prevail which are not reproducible on Earth; and, during the eclipse, the Sun's extended envelope, the corona, is accessible for observation and experimentation. Secondly, the study of the Sun is a natural and integral part of our study of the Universe: if we would understand the physics of the billions of stars scattered through space, we must first of all understand the Sun and its environment 'on our doorstep'. Finally, as we have already seen, Sun–Earth interactions directly affect all of us.

But what aspects of solar eclipses interest scientists, and what are their motivations for observing them?

Observation of the solar corona

On clear days the sky is too bright around the solar disk – even from mountainous regions – to allow observation of the whole corona in white light; but during a total

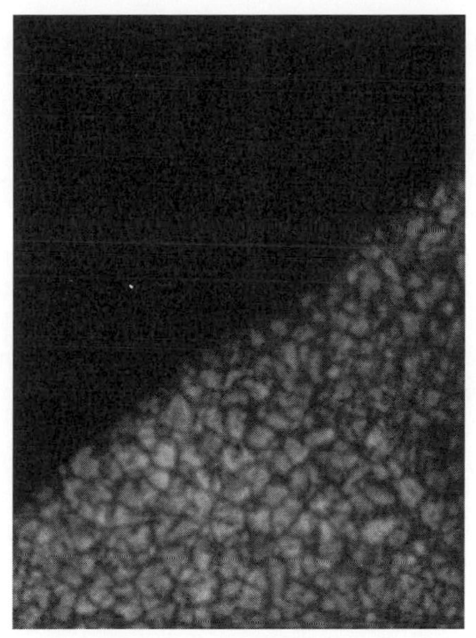

Fig. 3.5. A photograph in white light of solar granulation beyond the lunar limb, during a partial eclipse in 1973, taken with the 40-cm vacuum Newton telescope of the Fraunhofer Institute, at Izana, Tenerife. This type of observation can be used to correct images, due to the presence of the Moon's edge, which acts as a kind of Foucault 'knife-edge'. (F.L. Deubner, W. Mattig, Fraunhofer Institute.)

eclipse of the Sun, sky brightness is reduced by a factor of between 1,000 and 10,000. For observers in the Moon's shadow, sky brightness is due only to the scattering of light reflected by the ground and by the lower atmosphere in the area surrounding the shadow. This brightness will vary as the Moon's shadow passes rapidly across. During the eclipse, the sky will be dark enough for structures of the solar corona to be seen and studied.

Other possibilities for study

Other measurements, of less import and interest than those of the corona, can be taken during eclipses of the Sun.

Measuring the Sun's diameter Measuring the duration of the eclipse allows us to measure the Sun's diameter. Account must be taken of the profile of the lunar limb and the observer's position *vis-à-vis* the centre line, as corrections involving the relative positions of the lunar and solar disks will be needed if the observer is not on the centre line. Accuracy in the measurement of this interval must be to within 0.01 s, corresponding to 0.005 arcsec of the solar diameter, or 4 km in 1,391,000. Such measurements are used to verify hypothetical variations in the diameter of our star.

Measuring the shape of the Moon The Moon is an ellipsoid with three axes. The major axis points towards the Earth, while the other two are at right angles to this axis. Eclipses of the Sun have resulted in very precise measurements of the shape of the Moon, by observers stationed on either side of the centre line.

The progress of a total eclipse is marked by *contacts. First contact* is the moment when the Moon takes the first 'bite' out of the Sun's disc. *Second contact* is the disappearance of the last gleam of the Sun's light: a great moment for the observer, at the onset of totality. When totality ends, seconds or minutes later, the Sun reappears in a burst of light as *third contact* occurs. *Fourth contact* is the moment when the Moon moves completely off the Sun, which resumes its circular aspect.

The solar disk is considered to be circular, and it is possible to determine certain lunar dimensions by measuring the time elapsed between second and third contacts, thus increasing our knowledge of the Moon's geometry.

Refinement of ephemerides The precise instants of contacts, calculated by means of ephemerides, cannot be totally accurate. These inaccuracies, of a few seconds, are due to our lack of knowledge of the exact shape of the Moon, the diameter of the Sun, and other factors. Measuring the times of contacts can improve our ability to perform positional calculations.

Examples of solar eclipses: total, annular, partial

Each type of eclipse varies in its characteristics, such as maximum duration, diameter of the shadow, and definition of contacts.

We now discuss the differences, and the common and distinct parameters, characteristic of total, annular and partial eclipses.

Total eclipses

The maximum duration of the phase of totality for any total eclipse of the Sun is 7 m 58 s. A future eclipse which will almost attain this duration of totality is that of 5 July 2168, with 7 m 29 s. The medial line of the narrow strip swept by the Moon's shadow is called the centre line. If an observer is within the shadow track, but not on the centre line, the duration of totality will be reduced, and may be corrected for by using the formula

$$d = D \, . \, [1 - (2a/w)^2]^{1/2}$$

where d is the duration of totality for the observing site in seconds; D is the duration of totality on the centre line in seconds; w is the width of path of totality in kilometres; and a is the distance in kilometres from the observing site to the centre line.

If t_m is the value determined for the time of maximum eclipse for a given site, then the approximate times of second (t_2)and third (t_3) contacts will be

$$t_2 = t_m - d/2$$
$$t_3 = t_m + d/2$$

The longest possible duration for the lunar shadow-cone's journey across the Earth is 4 h 30 m in equatorial regions, and 3 h 30 m at latitude 45°.

The *magnitude* of an eclipse is the parameter defining the amount of the solar surface covered by the Moon. The magnitude of an annular or partial eclipse will always be less than 100%; but for a total eclipse the figure, from the zone of totality, will be equal to, or even slightly greater than, 100%, because the Moon can present a larger apparent diameter than the Sun during a total eclipse. If the magnitude of the eclipse is only a little more than 100%, the inner regions of the solar atmosphere, the chromosphere and the prominences, will be visible throughout totality. With magnitudes any larger than this – up to a maximum of 106.5% – the Moon will be large enough to hide the chromosphere and any prominences. These phenomena may, however, be observed just after second contact or just before third contact, at the eastern and western limbs respectively.

The Moon's disk is not perfectly regular at its edge. Just before second contact and just after third contact, valleys at the Moon's edge may allow through a tiny fraction of the light from the photosphere. This light, shining between the lunar heights, causes the phenomenon known as *Baily's Beads*. When the final 'bead' remains, set in the pink circle of the chromosphere, the *Diamond Ring* results.

In calculating durations of eclipses, a fundamental parameter is the mean lunar radius k, expressed in terms of the Earth's equatorial radius. The times of contacts, magnitudes and durations of totality all depend on the angular diameters and relative speeds of the Moon and the Sun. If the Moon were to present a perfect disk, such calculations would be simple. Unfortunately, the accuracy of our mathematical predictions is limited because of the irregular profile of the Moon's limb. The Moon's radius – a basic parameter for the calculation of eclipses – varies as a result of the irregularities at the limb, although calculations based on a mean radius, averaging out mountains and valleys, are often sufficiently accurate for most purposes. In August 1982, at a general meeting of the International Astronomical Union (IAU), a value for k of 0.2725076 was adopted. This is the best estimate, taking into account the heights of mountains and depths of valleys on the Moon's rugged edge are taken into account. Even so, problems can still arise when employing this mean IAU value for the prediction of the duration and, more especially, of the type of eclipse. Such was the case with the eclipse of 3 October 1986, which, according to the *Astronomical Almanac*, would be total for 3 seconds. In the event, Baily's Beads remained visible throughout totality. This example clearly shows the importance of knowing the Moon's radius and the profile of its limb.

Some specialists nowadays use the IAU value for k for the first and fourth contacts, but for the second and third contacts they use a lesser value to predict the duration of totality, and as a precaution against wrongly predicting the type of eclipse. If very great precision is required, the profile of the lunar limb must be taken into consideration. There is now an impressive amount of data on this, assembled by Watts in 1963, and completed from analyses of occultations of stars by the Moon in the 1970s by van Flandern, Morrison and Appleby.

This database of lunar topography can help to predict the appearance of the Sun's light at points around the lunar limb. For each eclipse, limb profiles are prepared and published by NASA and the Bureau des Longitudes. These profile diagrams (see Figure 3.6), so full of information, are worth exploring. The figure shows the lunar limb, much exaggerated, with peaks and valleys at the edge of the disk; the mean

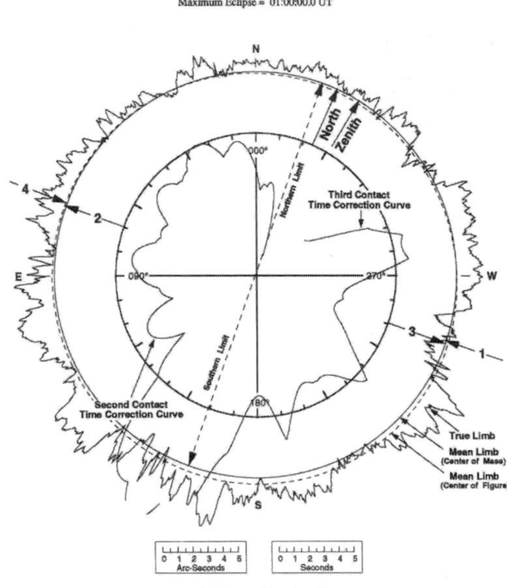

Fig. 3.6. Prediction of the positions of Baily's Beads is possible due to studies of the profile of the lunar limb. For example, on this diagram from the NASA bulletin on the eclipse of 9 March 1997, arrows 2 and 3 show the position of second and third contacts. The time correction curve for the second contact shows a delay in the time for the 180° position, and there will therefore be a 'bead' at this location. The time correction curve for third contact shows an earlier time at 250°, so the first rays of the photosphere will appear at this location. (F. Espenak, NASA/GSFC.)

limb, as a circle, with peaks and valleys 'ironed out'; and the mean limb based on the centre of mass, taking into account corrections for the effects of ellipticity and libration on the Moon's path. Also shown is a circle graduated in degrees from 0° to 360° for the purposes of orientation. Within this circle are two meandering curves, which show time corrections for the second and third contacts. Arrows 1, 2, 3 and 4 mark the location of the contacts.

By examining the relief at the limb and the time correction curves for the contacts, we can determine the locations where Baily's Beads will appear. Near the eastern edge, where second contact takes place, it will be noted that where the limb shows pronounced dips the time correction curve has humps. This means that, for those locations, second contact, at the onset of totality, will be a little later, and the photosphere will be seen for a little longer. For the third contact, on the western side, places where the limb shows dips are echoed by similar inward plunges of the time correction curve. This means that, for these locations, third contact, at the end of totality, will be slightly ahead of time, and here the photosphere will first be glimpsed.

The shadow zone on the Earth's surface will be delimited by the intersection of the umbral cone and that surface. The shadow will be elliptical in form, its eccentricity increasing with latitude. The maximum dimension of the shadow is 269 km for a total eclipse, although its projection onto the Earth's surface may be larger: the shadow zone of the eclipse of 5 May 2600 will be a maximum 627 km across.

Annular eclipses

During an annular eclipse the photosphere remains visible throughout the event, as the diameter of the Moon is insufficient to cover the Sun completely. The Sun will appear as a ring of light around the Moon.

The maximum duration for the 'total' phase of an annular eclipse of the Sun is 12 m 30 s. A forthcoming annular eclipse with a duration approaching this maximum value will be that of 15 January 2010, with the annular phase lasting 11 m 7 s. The maximum dimension at any time for the zone of visibility of an annular eclipse is 370 km.

The minimum possible coverage of the Sun by the Moon at mid-eclipse is 89.1%, which means that a ring, comprising 10.9% of the Sun's surface, will be seen. This phenomenon is sometimes known as the 'ring of fire'.

Second and third contacts have a different meaning in the case of an annular eclipse. Here, second contact is the moment when the ring phenomenon appears, with the Moon tangential to the right-hand edge of the Sun. Third contact is the moment when the ring ceases to be complete, with the Moon tangential to the left-hand edge of the Sun.

An eclipse may be both annular and total, when the Moon's shadow – at first not long enough to reach the Earth's surface, touches down during the event as a result of the spherical nature of the Earth. On 8 April 2005 there will be an annular eclipse, beginning in the Pacific Ocean, which becomes total as the shadow moves towards the Earth's equator.

Partial eclipses

During a partial eclipse, a fraction of the solar disk remains visible. There are therefore only two distinct periods in the sequence of a partial eclipse, and only two contacts which enclose the phenomenon. The first contact is the moment when the lunar disk encroaches upon the Sun. The second, and final, contact corresponds to the departure of the Moon from the Sun's disk, from which moment the event is over. The maximum phase of a partial eclipse occurs when the largest fraction of the Sun's disk is hidden.

The main parameter of interest with a partial eclipse is its *phase*; that is, the fraction of the Sun's disk hidden by the Moon. Another parameter, known as *magnitude*, is sometimes used; it is the percentage of the solar diameter covered by the Moon. The moment of maximum eclipse, and the percentages of the surface and the solar diameter hidden, vary with the location of the observer.

In the ephemerides, there is reference to a parameter known as *gamma*, expressed in Earth diameters. This defines the distance between the axis of the lunar shadow and the centre of the Earth. If gamma is equal to zero Earth diameters, then the axis of the shadow passes through the centre of the Earth. If gamma is positive (or

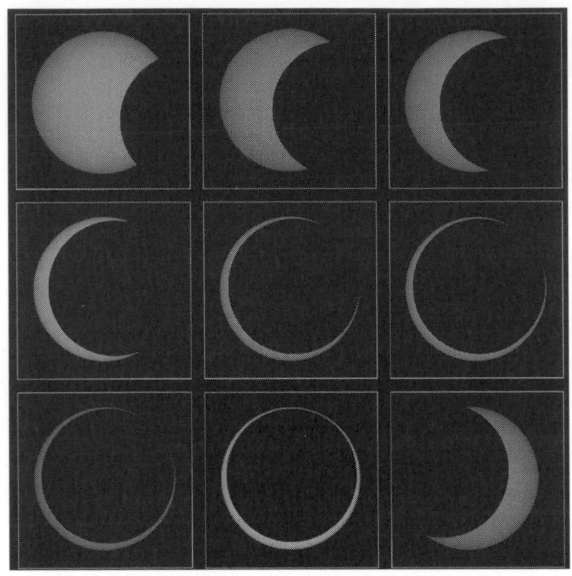

Fig. 3.7. On 10 May 1994 an annular eclipse was seen from the United States. This series of nine photographs shows the progress of the event. The photograph at bottom centre was taken at the moment of totality. Note that the observer was not exactly on the centre line, as the photospheric ring varies in width. (Richard Tresch Fienberg, *Sky and Telescope*.)

negative), the axis of the shadow passes north (or south) of the equator. If gamma is exactly equal to the diameter of the Earth, the axis of the shadow will be tangential to the Earth's north pole; and if it lies between 1.001 and 1.003 (or –1.001 and –1.003), this signifies that the axis of the shadow misses the Earth. However, a fraction of the shadow may touch the Earth, and the eclipse is then non-central total or non-central annular. If gamma is greater than 1.003, or less than –1.003, then the eclipse is partial.

The term 'non-central' is used when the axis of the Moon's shadow does not touch the Earth's surface. This type of eclipse occurs only near the poles. These eclipses are normally partial, but may on rare occasions be total or annular in these regions if a fraction of the shadow grazes the Earth.

ECLIPSES OF THE MOON

Eclipses of the Moon have excited major scientific interest throughout history. They have been used in the investigation of the dimensions of our natural satellite, and to measure the distance between the Earth and the Moon. These measurements can nowadays be repeated quite easily by anyone, using only a watch, while observing this magnificent spectacle of the night.

We now turn to the appearance, contemporary scientific interest and observation of lunar eclipses.

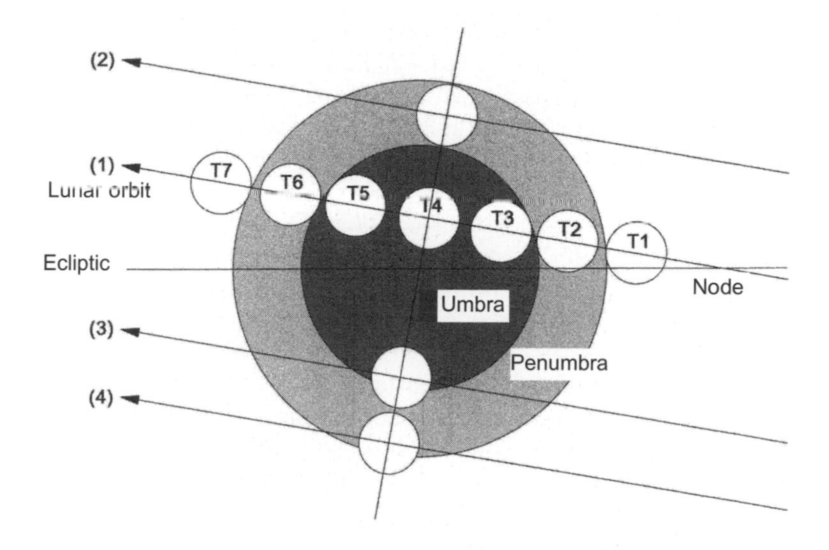

Fig. 3.8. Different circumstances for the passage of the Moon through the Earth's umbra and penumbra. Trajectory (1): total eclipse within the umbra. Trajectory (2): total penumbral eclipse. Trajectory (3): partial eclipse within the umbra. Trajectory (4): partial penumbral eclipse. T1 to T7 show the various contacts of the Moon's limb with the Earth's umbra or penumbra. (P. Guillermier.)

In the shadow of the Earth

Depending on the relative positions of the Moon and the Earth's shadow, an eclipse of the Moon may be:

- total, if the Moon enters the cone of the Earth's shadow completely;
- partial, if only a fraction of the Moon's surface enters the Earth's shadow;
- penumbral, if the Moon travels only through the Earth's penumbra.

Since the Earth's shadow points away from the Sun, an eclipse of the Moon can occur only if the Moon lies in this direction – at full Moon.

Because of our knowledge of the parallaxes of the Sun (9″) and the Moon (3′ 41″), and of the Sun's angular radius (16′), we may calculate the angular diameters of the umbra and the penumbra. The value for the umbra is 41′ 30″, and for the penumbra 1° 13′ 30″. Since the mean apparent diameter of the Moon is 31′, it is possible to demonstrate that our satellite can fit easily into the umbra and the penumbra.

As an eclipse proceeds, the eastern part of the Moon encroaches into the penumbra. This produces a diminution of light intensity, albeit slight, as seen by the observer, and lasts for about an hour. Then, as the Moon enters the umbra, a crescent-shaped part of the lunar surface is darkened. The shadow finally covers the whole disk, and, at totality, the Moon takes on a reddish hue. The duration of totality, depending upon the path of the Moon through the umbra, may last from just a few seconds to as long as 1 h 45 m. When totally eclipsed, the Moon does not disappear completely, but the reddish colour varies from eclipse to eclipse. It is caused by refraction of sunlight through the Earth's atmosphere, with blue light being absorbed and red light transmitted. The residual colour and brightness depend on the degree of transparency of the Earth's atmosphere (where the presence of volcanic aerosols is a factor), solar activity and the distance of the Moon from the Earth at the time.

If the Moon is at apogee (at its greatest distance from the Earth), its position is closest to the apex of the umbral cone, and the Moon's surface is normally brighter than if the Moon were at perigee, with the Moon further down towards the base of the cone. Refraction of the Sun's rays is considered to be negligible at an altitude greater than 75 km. For rays passing very close to the Earth's surface, deviation is at maximum, and these rays converge at a distance of about 260,000 km. Since the Moon is further away than this, at a distance of at least 356,400 km, it will never be totally eclipsed. The lunar surface will always receive some refracted light from the Earth's atmosphere. French astronomer André Danjon has drawn up a scale for estimating the brightness of total lunar eclipses. Estimates must be made at mid-eclipse, when the Moon is nearest to the centre of the umbral cone, with the naked eye or with a low-power instrument (magnification less than $\times 20$).

Table 3.2. Scale for estimating the brightness of lunar eclipses. (André Danjon.)

Scale	Appearance of the Moon at mid-eclipse
0	Very dark eclipse. Moon hardly visible, or invisible.
1	Dark eclipse, grey or maroon. Lunar features difficult to see.
2	Dark or rust-coloured eclipse. Centre of umbra very dark, with brighter periphery.
3	Brick-red eclipse. Edge of umbra has greyish-yellow tint.
4	Bright eclipse, orange or copper-coloured. Edges bluish and bright.

During the eclipse of 19 March 1848, the Moon remained so bright that observers wondered if an eclipse was actually happening. At the other end of the scale, the Moon was invisible to the naked eye at mid-eclipse on 18 May 1761, and on 10 June 1816. Analysing measurements on the basis of his scale, Danjon showed a relationship between brightness in the umbra and the 11-year cycle of solar activity: at solar maximum, the lunar surface shows brighter hues.

The significance of lunar eclipses

By observing eclipses of the Moon – or, more precisely the shape of the shadow cast by the Earth upon the lunar surface – it can be deduced that the Earth is a sphere. In the third century BC, Aristarchus of Samos calculated the diameter of the Moon by measuring how long it took to pass through the Earth's shadow. He arrived at a reasonable estimate of 4,600 km, the actual value being 3,476 km. Looking back at Eratosthenes' calculation of the size of the Earth, Aristarchus applied a

Fig. 3.9. A photograph of the lunar eclipse of 3/4 April 1996. The Moon, here in mid-eclipse, was photographed at 0h10m UT on 4 April through a 180-mm Takahashi Mewlon reflector, of focal length 1,300 mm, using Fuji 400 ISO film and an exposure of 5 s. (P. Guillermier.)

trigonometric method to his result and calculated the distance from the Earth to the Moon. Hipparchus refined these estimates in about 150 BC, and assigned values of 4,200 km for the diameter of the Moon, and 425,000 km for its distance. It was left to Ptolemy, in the second century AD, to calculate values of astonishing accuracy: 3,700 km for the lunar diameter, and 376,000 km for the distance.

In the seventeenth century, a partial solution to the urgent problem of the determination of longitude involved eclipses of the Moon, which could be observed from many places simultaneously; and because of these measurements, in 1634 the length of the Mediterranean Sea was found to be 1,000 km less than had previously been thought.

By studying light falling upon the Moon during eclipses, Daniel Barbier and Daniel Chalonge of the Institut d'Astrophysique in Paris discovered that part of the Earth's ozone layer – so vital for the existence of life on our planet – is confined to a level between 50 and 80 km in altitude. Another modern development, which takes advantage of the absence of reflected sunlight during eclipses, is the use of lasers, pointed at mirrors left on the Moon during the Apollo and Lunakhod missions, to make precise measurements of the Moon's secular acceleration and the slowing of the Earth's rotation. Astronomers at the Lick Observatory first made such measurements in 1969, but not without difficulty, due to a 545-m error in the position of the observatory. Lunakhod 2 left behind a laser reflector, the TL2, with 14 cataphotes, and the Apollo 11 and Apollo 14 missions deposited two 100-element reflectors in the Mare Tranquillitatis and at Fra Mauro. The Apollo 15 mission set up a laser reflection panel with 300 elements near the Hadley Rille site, and laser measurements were carried out by the McDonald Observatory in Texas, in conjunction with instruments at Haleakala in Hawaii, and at Cerga, near Grasse in France. Since we know that the speed of light in a vacuum is 299,792,458 m s^{-1} – a value used as the basis for the standard metre – light-travel time to and from the Moon can be determined to an accuracy of a few tens of picoseconds by the laser method, using a caesium clock. This interval, of about 2.5 seconds, defines the distance to the Moon to within 10 mm. As a result of such experiments, we now know that the Moon is escaping the Earth's gravitational attraction by one or two metres per century. What is measured here is the segment between a point on the Earth and a point on the Moon, and the value for this distance is determined not only by the distance to the Moon as it pursues its eccentric orbit; it is also dependant on lunar libration, the plasticity of the Moon, the influence of the Sun and the rotation and tides of the Earth. Thus it is that the compilation of data has led to a rich harvest of astronomical information.

Observing a lunar eclipse

Observation of the Moon's entering and leaving the penumbra is often neglected, because visually the Moon appears unchanged at these times. It may be, however, that the intensity of the penumbra will be greater than predicted.

More remarkable are the phenomena observed as the Moon enters and leaves the umbra. Aspects of the umbra include:

- the edge of the umbra (whether it is sharp or diffuse);
- the shape of the umbra: its edge is normally circular, but irregularities due to optical and topographical effects are often reported;
- density and colouration: these parameters are difficult to determine, since the brightness of that part of the Moon still illuminated interferes with estimates, being 100–1,000 times greater than that of the shadowed area.

During totality, neither the intensity of the umbra, nor its colour, are homogeneous across the eclipsed disk of the Moon. These variations are caused by refraction of sunlight into the umbral cone. It has been shown that there exists a phenomenon of focalisation in the centre of the umbra, and nuances of intensity and colouration may also be ascribed to variations in atmospheric conditions at the Earth's terminator. Because of the Earth's rotation, there may be considerable dissimilarities in the appearance of the eclipse between the first and second halves of totality.

It is also interesting to observe the visibility of certain lunar formations within the umbra; for example, the brightest features, such as Aristarchus, Kepler and the rays of Tycho, and the darkest, such as the maria and certain craters.

Transient lunar phenomena

Very occasionally, temporary anomalies – such as increases in brightness or apparent reductions in albedo – are observed on the Moon. These peculiar events are known as transient lunar phenomena (TLP). The first identification of such a phenomenon was by astronomer Dinsmore Alter on 26 October 1956. On one of his photographs, the crater Alphonsus appeared to be outgassing a cloud. Following this discovery, Russian astronomer Nikolai Kozyrev began a systematic study of this region with the 1.25-m telescope of the Crimean Astrophysical Observatory. On 3 November 1958 the phenomenon reappeared and, with the spectrograph attached to the instrument, Kozyrev was able to analyse the burst of gas. Carbon vapour in particular was detected. The spectrum obtained remains one of the rare pieces of evidence in favour of the existence of TLP. During an eclipse of the Moon, the particular conditions prevailing at the surface may favour observation of such apparitions.

There are various kinds of TLP:

- unusual increases in brightness over a restricted area of the Moon;
- decreases in brightness;
- colour effects, the majority reported involving red, or sometimes blue hues, and more rarely orange, green or yellow;
- fog-like effects, with shimmering areas in otherwise sharply defined surroundings. This effect should not be confused with turbulence in the Earth's atmosphere.

Some TLP combine various aspects. They may last from 0.5 s to 30 m, and reappear intermittently for several hours. The area affected is about 5–15 km across. Whatever process or processes give rise to such observations is still not understood. Theories have included rapid emission of gases through the surface, and drifts of dust, whose particle size is heterogeneous, capable of emitting light through triboelectric effects.

TLP are not randomly scattered about the lunar surface: they concentrate near the edges of maria (north of the Mare Imbrium, south of the Mare Serenitatis and around the Mare Crisium), and at certain craters. The appearance of TLP might therefore be linked to lines of weakness in the lunar crust. During an eclipse of the Moon, the amount of energy received at the lunar surface is suddenly reduced, and the resulting rapid variation in temperature may well induce thermal constraints which may shatter rocks, liberating trapped gases from cavities. Potential sites for TLP (and well worth watching) include Aristarchus, Alphonsus, Copernicus, Gassendi, Plato, Proclus, Theophilus, Menelaus, Herodotus, Eratosthenes, Grimaldi, Stöfler and Tycho.

Measurement of the Moon's distance during an eclipse

Before undertaking this measurement, the tapering of the Earth's umbral cone at the distance of the Moon must be evaluated. This information can be derived from observation of a total eclipse of the Sun. During this event, the Moon's shadow falls on the Earth, and we notice that the shadow zone is quite small. During an annular eclipse, the umbral cone of the Moon comes to a point in slightly less than the distance from the Moon to the Earth. We may therefore deduce that the diameter of the Moon's umbral cone decreases by one lunar diameter over the Earth–Moon distance.

Another datum needed for calculating the Earth–Moon distance is the apparent

diameter of the Moon. This may easily be done with the aid of a small disk (for example, a 5p coin) placed at a distance such that it will just cover the Moon. The 5p coin, 18 mm in diameter, has to be at a distance of 2.07 m in order to do this. The ratio of diameter to distance is 1:115 (an angle of half a degree in the sky), which shows by proportionality that the distance from the Earth to the Moon is 115 times the lunar diameter. Finally, the speed of the Moon's motion around the Earth can be estimated. One lunation lasts 30 × 24 hours (720 hours), and the length of the path is $0.5°/360° = 720$ lunar diameters. Therefore, the Moon moves a distance close to its own diameter during 1 hour of its orbit.
If:

T_1	:	beginning of entry into the umbra
T_2	:	beginning of totality
T_3	:	end of totality
T_4	:	exit from the umbra
R_{umbra}	:	radius of the Earth's umbra
R_{Moon}	:	radius of the Moon
R_{Earth}	:	radius of the Earth

then

$$R_{umbra}/R_{Moon} = (T_2 - T_1) \,.\, (T_4 - T_2)$$

or:

$$R_{umbra}/R_{Moon} = (T_4 - T_3) \,.\, (T_4 - T_2)$$

As a numerical example, we may take the values for the total eclipses of the Moon on 21 January 2000 and 16 July 2000.

For the eclipse of 21 January 2000, the values are

T_1 = 3 h 2 m
T_2 = 4 h 5 m
T_3 = 5 h 22 m
T_4 = 6 h 24 m

$$R_{umbra}/R_{Moon} = 2.43$$

The eclipse of 16 July 2000 gives:

T_1 = 11 h 58 m
T_2 = 13 h 2 m
T_3 = 14 h 48 m
T_4 = 15 h 53 m

$$R_{umbra}/R_{Moon} = 3.04$$

The mean ratio is

$$R_{umbra}/R_{Moon} = 2.74$$

Now, since

$$R_{Earth} = R_{umbra} + R_{Moon}$$

then

$$R_{Earth} = 3.74 \, . \, R_{Moon}$$

We know that the radius of the Earth, as measured by the method of umbral distances, is 6,378 km, so it follows that

$$R_{Moon} = 1{,}705 \text{ km}$$

The true value for the Moon's radius is 1,738 km.

The distance from the Earth to the Moon can also be deduced:

$$\text{Earth–Moon distance} = 115 \, . \, D_{Moon}$$

Therefore

$$\text{Earth–Moon distance} = 392{,}150 \text{ km}$$

This measurement does not take into account the Moon's elliptical path around the Earth. However, the result is quite close to the real value, which varies from 356,400 km at perigee to 406,704 km at apogee.

ECLIPSE CALCULATION PROGRAM

The power of modern microcomputers, and even programmable calculators, now presents amateur astronomers with the opportunity to run various simulation programs in conjunction with their observations. Involving as it does precise and repeated calculations, and the possibility of verification of certain conditions, computer work lends itself to astronomy.

To write the solar and lunar eclipse program (Appendix C), it was necessary to determine the conditions which must be fulfilled by the Sun, the Moon and the Earth for an eclipse to occur:

1) The longitudes of the Sun and the Moon must be equal (eclipse of the Sun) or 180° apart (eclipse of the Moon);
2) The longitudes of the Moon and one of the nodes of its orbit must coincide.

These two conditions must be fulfilled simultaneously. If they coincide only approximately, then a partial eclipse will result.

It would be unprofitable here to try to work out precise values for the sizes of the umbra and the penumbra during the eclipse, by calculating the relative positions of the Sun and the Moon with respect to the Earth at every moment; so the elements used for calculating positions in this program are insufficiently accurate. They can be used to predict only the dates of eclipses, and not to locate the zones of the Earth's surface affected.

To achieve our aim, we have to calculate, for every day in one year:

- the Sun's geocentric longitude, S
- the Moon's geocentric longitude, M
- the longitude of the ascending node, N

These values are calculated using mean positional elements obtainable from various publications (see Bibliography).

The conditions for the proximity of the Moon to one of its nodes are different for lunar and solar eclipses, because of the difference between the lunar and terrestrial diameters. If the longitude of the node is close enough to the Moon's longitude; that is, if

$M - N \leq 11°$ for eclipses of the Sun

and if

$M - N \leq 15°$ for eclipses of the Moon

then an eclipse may occur, as long as the longitude of the Sun, S, is equal to the longitude of the Moon, M, to within 180°. ($S - M = 0° + k \, . \, 180°$, where k is a whole number).

Moreover, the Moon's longitude evolves 12.36 times faster than that of the Sun. This means that 29.53 days (one synodic month) elapse between two successive passages of the Moon before the Sun, and the event occurs 12.36 times per year.

The two longitudes S and M may be equal, or 180° apart. Only some of the 25 possibilities may satisfy the extra condition of the Moon's being near the node. These few rare episodes will bring about eclipses.

In the program, calculations are carried out day by day, at 0 hours. As coincident events occur most often between two consecutively calculated daily values, the time of the coincidence is arrived at through linear interpolation.

For eclipses of the Sun, the program calculates differences DM and DN, where

$DM = S - M$ (difference between longitudes of Sun and Moon)

and

$DN = S - N$ (differences between longitudes of the Sun and node)

When, from one day to the next, DM reduces to zero, the conditions

$DN \leq 11°$

and

$DN + 180° \leq 11°$

are examined.

If one of the aforementioned two conditions is fulfilled, an eclipse of the Sun will occur, and the time of the coincidence is sought.

For eclipses of the Moon, the program calculates differences DM and DN, where

$DM = S - M + 180°$ (difference between longitudes of Sun and Moon to 180°)

and

$DN = S - N$ (differences between longitudes of Sun and node)

If DM reduces to zero, the conditions

$DN \leq 15°$

and

$DN + 180° \leq 15°$

are examined.

Here too, if one of the two conditions is fulfilled an eclipse of the Moon will occur, and the time of the coincidence is sought.

After searching through and calculating for every day of the selected year, the program displays the type of eclipse (solar or lunar), the number of the day in the year, and the time of the event.

This is a simple program, and, in the case of an eclipse of the Sun, it will not provide the path of the shadow zone across the Earth. It is, however, possible to make a rough estimate of the region of the globe involved, which will depend on the position of the node in relation to the Moon at the moment of conjunction:

- If the ascending node of the Moon is situated after the conjunction, the Moon's umbra will be projected onto the southern hemisphere. The same will be true if the descending node is situated before the conjunction.
- If the ascending node is situated before the conjunction, the Moon's umbra will be projected onto the northern hemisphere, and the same will be true if the descending node is situated after the conjunction.

The eclipse calculation program, detailed in Appendix C, is written in BASIC, but transcription into other languages such as FORTRAN, PASCAL or C should present no difficulties.

4

Historical eclipses and discoveries

The first dated observations of these celestial phenomena come from Indian, Chaldean, Babylonian and Chinese sources. The demands of astrology and the calendar led to the first human efforts to predict eclipses.

This activity required painstaking study of former observations. Thus, in the Library of Assurbanipal at Nineveh (650 BC) were found large numbers of astronomical tablets, the most ancient dating from the twentieth century BC, giving details of past eclipses and attempting to predict others.

Eclipses of the Sun and Moon have marked out the course of human history, and have sometimes even changed it. We shall now look at some of the best known legends and episodes concerning eclipses throughout history, and then touch on some remarkable landmark experiments in solar physics.

MYTHS AND LEGENDS

Ancient myths and legends which attempt to explain eclipses are present in all cultures. Most of them tell of creatures devouring the Sun during some celestial repast. The Indonesians and the Chinese believed that an enormous dragon ate the Sun – a concept also found in Greek and Roman legend, and later among the Balts. The Mandarin word for eclipse is *shi*, which literally means 'eat'.

This theme of supernatural beasts devouring the Sun and Moon during their periodic disappearances recurs in many cultures, but the beast in question differs. In Siberia, the culprit was a vampire; in Serbia, a werewolf; in Vietnam, a giant frog; in Paraguay and Argentina, a jaguar; and in Bolivia, a huge dog.

According to the Vietnamese legend, eclipses of the Moon were the result of an enormous frog determined to swallow it. The giant amphibian, tied to the pool of Hanh by a golden chain, sometimes tried to escape when the neighbouring lord was sleeping. The ladies of the Moon would then rush to wake the lord of Hanh, who alone was able to make the frog disgorge the Moon. Young girls still beat their rice-grinding bowls with pestles to help the ladies wake the lord.

In India, two dragons, one at each node, were held responsible for the

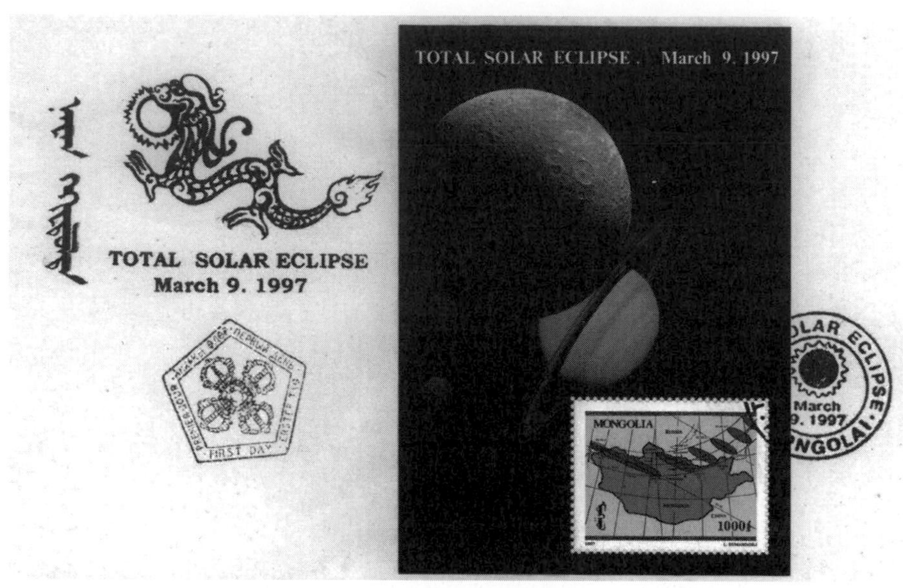

Fig. 4.1. Mongolian first-day cover commemorating the solar eclipse of 9 March 1997. The motif of a dragon eating the Sun recalls the legend. (Collection of P. Guillermier.)

phenomenon. Here is the origin of the term draconic, used to designate the period between two passages of the Moon at the nodes of its orbit.

One of the most interesting legends is told on the island of Bali, in Indonesia. It was inspired by the Hindu epic, the Mahabharata. The malevolent Kala Rau (Rahu in the Indian version) is jealous of the immortal and omniscient gods inhabiting Nirvana. Kala Rau lays his plans to achieve immortality. Disguised as a woman, he connives to be present at the gods' banquet, serving their magical elixir. Taking advantage of a disturbance, he furtively takes a mouthful of the drink, but Vishnu, aware of the crime, cuts off his head immediately he has committed it. Kala Rau's decapitated body dies, but his head has been made immortal by the potion. Ever since, the head has chased the Moon and the Sun through the sky in an attempt to catch and eat them, but when it succeeds they reappear, after a brief absence, through his open throat.

In the same ancient Hindu text – the Mahabharatra – there is a reference to an eclipse of the Sun, referring to the great battle of Kurukshetra, fought on the day of the eclipse, when the supreme god Vishnu revealed himself.

The most ancient eclipse recorded seems to be an eclipse of the Sun mentioned in an Indian chronicle. The event took place on 21 October 3784 BC.

Looking back into the past, we find that the Chinese were certainly capable of predicting eclipses of the Sun and Moon as early as 2800 BC, due to their knowledge of the saros. The oldest Chinese text in which an eclipse figures is inscribed on bone, and was discovered at Anyang, in Hunan province. It dates from 1300 BC. Also from China comes the story of two luckless royal astrologers, Hsi and Ho, whose

Fig. 4.2. This caricature appeared in a special eclipse edition of an Indonesian newspaper on 11 June 1983. With the solar corona at a temperature of 2 million degrees, the dragon burns himself and spits out the orb of day. (Collection of S. Koutchmy.)

task it was to scrutinise the sky for signs and portents. They fell down on their duty, however, when they failed to predict an eclipse of the Sun, and were beheaded for their lack of alertness. Most astronomers have heard of this legend, and have taken to heart its lesson: instead of relying on mere speculation about future events, they have solved the problems of eclipse calculation in unbelievably vast detail. The astronomers of the Chinese court were given the task of organising demonstrations during eclipses, with archers firing arrows in the direction of the Sun, and drummers trying to frighten the dragon away. The story of Hsi and Ho is thought to date from between 2165 and 1948 BC, a time when knowledge of celestial mechanics among the Chinese was still too imprecise to allow proper calculation. A total eclipse of the Sun is recorded in the treatise on astronomy of Sung-shu, part of the official history of the Liu-Sung dynasty: 'Period of the reign Xio-qian, first year, seventh month, day bing-shen, first day of month. The Sun was totally eclipsed and the constellations shone'. The date given corresponds to 10 August 457 AD. At the time when this observation was made, the capital was Jian-kang (now Nan-ching).

For their part, Babylonian astronomers certainly observed the lunar eclipse of 2283 BC, and knew the periodicity of the saros, a development which accelerated progress in the science of celestial mechanics. One of the oldest texts, on a clay tablet bearing a description of an eclipse of the Sun, was discovered during excavations of the ancient Mesopotamian town of Ugarit, near the modern-day Syrian border. The text describes the eclipse of 3 May 1375 BC, and reads: 'On a day of the new Moon, in the month of Hiyar, the Sun was ashamed, and hid itself in the daytime, with Mars as witness.' Around 750 BC, Babylonian astronomers were systematically observing and cataloguing eclipses of the Sun and the Moon, together with many other celestial events. Only about 1,000 fragments of clay tablets, covered with cuneiform characters, have survived, and are mostly to be found in the British Museum. It seems that the Babylonian astronomers did not independently discover the cause of eclipses. They probably obtained this knowledge from the Greeks during the Hellenistic period in Babylon, in the late fourth century BC. Nevertheless, their eclipse observations are second to none in the ancient world. The Babylonian archives cover 50 eclipses of the Moon, from 700 BC to 50 BC. A few reports of complementary observations, compiled by Ptolemy in about 150 AD, can be found in his *Almagest*. Certain observations dating from 720 to 380 BC are obviously based upon Babylonian texts. On two clay tablets, one of which is badly damaged, while the other is perfectly preserved, there are two accounts of observations. They are both about the same eclipse, that of 15 April 136 BC: 'Year 175 (Seleucid), month 12, day 29, the eclipse began in the morning and became total. Mercury, Venus and other stars were seen. Jupiter and Venus, which were in their periods of invisibility, appeared during the eclipse.' Other and more recent clay tablets – part of an ancient Chaldean astronomical almanac – mention that an eclipse of the Sun and an eclipse of the Moon occurred in the same month. The following is an extract from the deciphered and translated text: 'Mercury gains height ... On the night of the 15th, forty minutes after sunset, an eclipse of the Moon begins ... Equinox ... On the 28th day, an eclipse of the Sun occurs.' In this extract, the dates are according to the Chaldean calendar. The eclipses were certainly those of 21 October and 4 November 425 BC.

With regard to Ancient Greece, there is a reference in Homer's *Odyssey* to a total eclipse near Ithaca on 16 April 1178 BC. Ptolemy reports observations of two eclipses in 720 and 721 BC. Thucydides, in his *History of the Peloponnesian War*, writes that the lunar eclipse of 23 August 413 BC convinced the Athenian general Nicias to postpone the evacuation of his army from Sicily for one month, during which time the Spartan army manoeuvred itself into position to block their escape route, destroyed the Athenian fleet and inflicted a major defeat upon the army of Athens.

Between 800 and 1100 AD, Arab astronomers, observing mostly from Cairo and Baghdad, described numerous eclipses of both Sun and Moon. Baghdad chronicler Ibn Al-Jawsi describes the event of 20 June 1061 AD thus: 'On Wednesday, two days before the end of the month Jumada Al-Ula, two hours after daybreak, the Sun was totally eclipsed. Darkness fell, and birds flew away. After more than four hours, the Sun appeared entire again. The eclipse was total only in the province of Baghdad.' We have shown that, thanks to eclipses, we can assign a place and a date to a great number of events, even though they have occurred far back in the past.

ECLIPSES OF THE MOON IN HISTORY

Eclipses of the Moon have made their influence felt in the course of human history. On 4 July 1917, Lawrence of Arabia succeeded in wresting the port of Aqaba from the Turks. Taking advantage of the darkness caused by an eclipse of the Moon, Lawrence and his Arab allies attacked from out of the desert while the town's defences faced mostly towards the Red Sea.

Other eclipses of the Moon have come down to us through history.

The fall of Constantinople

In 324 AD, the Roman emperor Constantine moved his capital to Byzantium, which he renamed Constantinople. For more than 1,000 years, the Byzantines governed the eastern Mediterranean and the Black Sea, ensuring the continuity of the Roman Empire. By the fifteenth century, dynastic strife, civil war, the attacks of the Crusaders, paralysing taxes and plagues such as the Black Death had weakened the Empire. As Constantinople foundered, the neighbouring Turkish sultanate plotted its overthrow. The Turks laid siege to Constantinople in 1402 and 1422, but failed to breach its formidable walls. However, in 1451 the young sultan Mohammed II assumed the Turkish throne, and planned a new assault. A cannon, invented by a Hungarian engineer and rejected by Constantinople for lack of funds, lay at the heart of his plans. This enormous gun, 8.2 m long, could fire 600-kg stones to breach the thick walls of the city. The sultan gathered an army of a quarter of a million men, for which the 7,000 soldiers in the city were no match. In April 1453, the Turkish army bombarded the walls of Constantinople, laying siege to the city. The defenders beat off three major attacks, and were able to repair the damaged walls by night. However, in spite of their defences, the Byzantines, heavily outnumbered, were worn down by the repeated assaults and constant repair work. At this stage in the siege, the morale of the defenders was buoyed up by an ancient prophecy, declaring that Constantinople could not fall if the Moon were waxing. But on 22 May the full Moon went into eclipse, dashing the morale of the troops. Contemporary accounts depart from reality, reporting that the Moon was in shadow for three hours. Six days after the event, Mohammed unleashed the final assault, and a gate fell at the height of the battle. Some of the Turks were able to penetrate the defences, and the battle turned into a rout for the Byzantines. The pillaging of the city lasted for three days. The rapidity with which the city fell indicates the general defeatism of the defenders, partly engendered by the eclipse of the Moon. The fall of Constantinople reverberated throughout Western civilization, and its influence on the development of European history lasted for centuries. (Table 4.1 provides details of the eclipse.)

Table 4.1. The lunar eclipse of 22 May 1453

Entry into penumbra	14 h 36 mUT
Entry into umbra	15 h 45 mUT
Exit from umbra	18 h 41 mUT
Exit from penumbra	19 h 50 mUT
Maximum phase	0.737
Circumstances for Constantinople (Istanbul):	
Moonrise	17 h 31 m UT
Local time	UT + 2 hours

Fig. 4.3. In 1453 the troops of sultan Mohammed II entered Constantinople. In spite of their strategic advantages, the Byzantines were outnumbered, poorly armed and demoralised as a result of an eclipse of the Moon. (Old engraving.)

Columbus's lunar eclipse

Eclipses have influenced more than the outcome of battles. During his voyages to the New World, Christopher Columbus used them for purposes both scientific and non-scientific, and was among the first to use an eclipse to measure the latitude of the observing site.

While exploring the coast of Central America during his fifth voyage, Columbus's exhausted expedition arrived in the West Indies. The tiny flotilla, worm-eaten and leaking badly, made landfall in Jamaica. Not long afterwards, half his men mutinied, stealing the food reserves, killing natives and engaging in guerrilla warfare against the crew members who remained loyal. Fearful of the mutineers, and unable to parley with Columbus's crew, the local inhabitants ceased providing food. Three days before an eclipse of the Moon, Columbus predicted to their chiefs that the god of the Christians, furious with the locals, would invoke a celestial sign of his displeasure. In a cloudless sky on the night of 29 February 1504, a flame-red Moon struck terror into the natives, who then supplied and protected the navigator and his party until a relief vessel arrived. (Table 4.2 provides details of the eclipse.)

Columbus's exploitation of the natives' ignorance of the mechanics of an eclipse has inspired other authors, such as Mark Twain (*A Connecticut Yankee in King Arthur's Court*) and Hergé (*Prisoners of the Sun*). In both books, the hero is saved from execution by an eclipse of the Sun.

Fig. 4.4. During his fifth voyage to the New World in 1504, Christopher Columbus predicted the disappearance of the Moon to frighten the Jamaicans, who were refusing to supply his men. The natives, terrified by the vision of the red, eclipsed Moon, subsequently helped the explorer and his party until relief arrived. (Illustration from *l'Histoire des Astres* by S. Rambooson.)

Table 4.2. The lunar eclipse of 29 February 1504

Entry into penumbra	21 h 53 m UT
Entry into umbra	22 h 59 m UT
Onset of totality	00 h 18 m UT
Maximum	00 h 41 m UT
End of totality	01 h 06 m UT
Exit from umbra	02 h 23 m UT
Exit from penumbra	03 h 29 m UT
Maximum phase	1.090
Circumstances for the West Indies (Havana, Cuba):	
Moonrise	18 h 34 m UT
Local Time	5 h UT

ECLIPSES OF THE SUN IN HISTORY

The Greek mathematician and philosopher Thales is said to have predicted, using the saros, the eclipse of 28 May 585 BC. He recounted that the shadow swept across the field of battle where the armies of King Alyattes of Lydia opposed those of King Cyaxarus of the Medes. As darkness came, the battle came to a halt, and a peace accord was signed. Since Thales' details were not very precise, the story has come down to us through Plato and the Christian writers.

Since those times, such fears have been somewhat allayed, and, thanks to eclipses, science has enjoyed a chain of discoveries.

Eclipses in literature

Literature is rich in references to eclipses. The Greek comic poet Aristophanes described an eclipse seen from Athens in his satire against Socrates, *The Clouds*. In his treatise *De Facie in Orbe Lunae* (*Of the Face in the Orb of the Moon*), Plutarch writes a dialogue between learned men, revolving mostly around the possibility of the Moon being inhabited. A reference in the work to an eclipse of the Sun has helped to date it to between 75 and 83 AD.

In his famous novel *King Solomon's Mines* (1885), H. Rider Haggard describes an eclipse of the Sun at full Moon. Apprised of the error, the author reworked the second edition, and the eclipse of the Sun became an eclipse of the Moon.

William Shakespeare refers to the lunar eclipse of 27 September 1605, and to the solar eclipse of 12 October 1605, in Act 1, scene 2 of *King Lear*. Gloucester attributes the ills of the kingdom to these celestial happenings ('these late eclipses in the Sun and Moon portend no good to us').

We also recall Mark Twain's *A Connecticut Yankee in King Arthur's Court*, and Blaise Pascal's *Pensées*; Pascal mentions the eclipse of the Sun of 12 August 1654.

We cannot list here all the literary references to eclipses, so we shall touch upon three very different subjects. In each case, an eclipse helps to locate the events in time or place.

The eclipse of Christ's Crucifixion

Historians have put forward many dates for the Crucifixion, relying upon the Hebrew calendar in use in Jerusalem during the first century.

According to the Gospels, Pontius Pilate, who passed sentence of death on Jesus according to the wishes of the sanhedrin, was procurator of Judaea at the time of the Crucifixion. This limits the event to between 26 and 36 AD.

Recent research into the Hebrew calendar – which is related to the Chaldean calendar – and comparison with the Julian and Gregorian calendars, indicates that Christ was crucified on Friday, 14 or 15 of the month of Nisan (corresponding to our April). As for the year, nearly all the indications favour 30 AD, although the year 33 cannot be completely ruled out. There is still some uncertainty as to the duration of Christ's public ministry, which may have gone on for two years and a few months, or three years and a few months. If the former is true, then Christ's death would have occurred in 29 AD. So, with this information in mind, historians suggest three possibilities: 29, 30 and 33 AD.

In the light of the scriptures, can astronomers offer precise estimates of the date? There is a reference in the Bible to darkness falling upon the place of the Crucifixion. In the Gospel according to St Matthew, the chapter telling of Jesus' death reads:

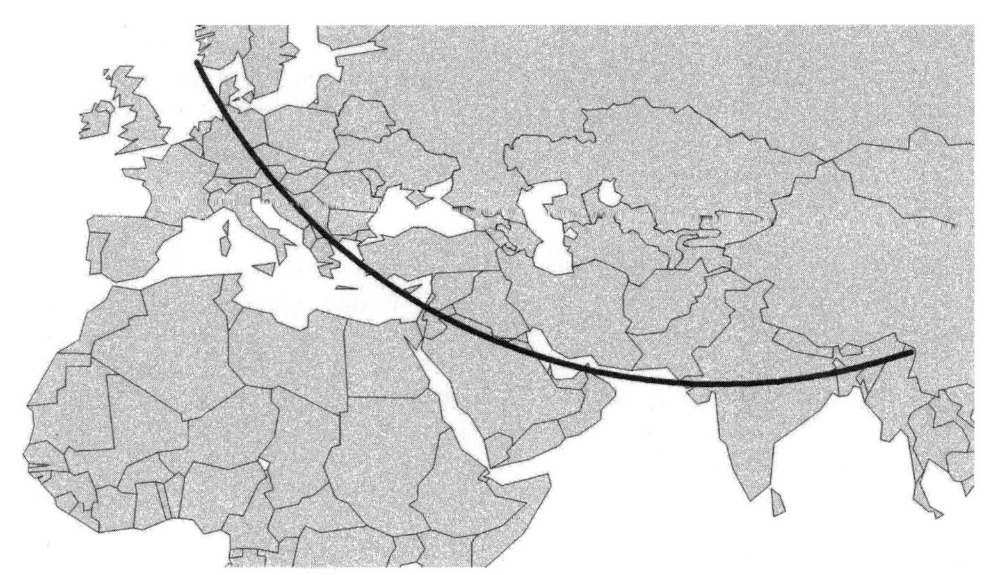

Fig. 4.5. The path of the eclipse of 24 November 29 AD, which preceded the Crucifixion. (P. Guillermier.)

'From the sixth hour until the ninth hour, darkness came over all the land.' In our system, these times correspond to midday and three o'clock in the afternoon.

The Acts of the Apostles describe how, on the day of Pentecost, Peter addressed the crowd, quoting the prophet Joel, in these words: 'In the last days, God says, I will pour out my spirit on all people . . . I will show wonders in the heaven above, and signs on the Earth below . . . The Sun will be turned into darkness, and the Moon to blood.' Calculations by astronomers have shown that there was indeed an eclipse of the Sun, visible from Jerusalem, but this took place during the morning of 24 November 29 AD. (Table 4.3 provides details of the eclipse.)

Table 4.3. The solar eclipse of 24 November 29 AD

First contact	07 h 35 m UT
Maximum	08 h 58 m UT
Last contact	10 h 27 m UT
Maximum phase	0.953
Local time for Jerusalem:	UT + 2 hours

In the book of Jeremiah, in the chapter headed 'disasters from the north', we find: 'I looked at . . . the heavens, and their light was gone.' In the book of the shepherd Amos, in the chapter concerning the wrath to come, reference to an eclipse is even more obvious: 'I will make the Sun go down at noon, and darken the Earth in broad daylight.'

Revelation, chapter 1, dealing with prophetic visions and the great day of judgment, contains another reference to an eclipse: 'I watched as he opened the sixth seal. There was a great earthquake. The Sun turned black like sackcloth made of goat hair, and the whole Moon turned blood red.' This description echoes closely what is seen and felt during an eclipse of the Sun, or, to be more precise, during an annular–total eclipse. When totality comes, the Sun appears black while hidden behind the Moon. During an annular–total eclipse, the chromosphere, and sometimes prominences, can appear surprisingly red. Such a phenomenon may well give the impression of a surge of blood.

In addition to biblical evidence, legend has it that a man called Dionysius observed an eclipse of the Sun at the time of Christ's Crucifixion.

The writings and accounts which have come down to us are sketchy and enigmatic, and we must exercise caution in their interpretation. Even though our calculations confirm that, obeying the dictates of celestial mechanics, the Moon's umbral cone did indeed sweep across the Earth's surface near Jerusalem on 24 November 29 AD at 8 h 58 m UT, it is important in our era, nearly 2,000 years later, to be circumspect and rigorous in dating with certainty an event of such world-historical importance. Nevertheless, it does seem rather unlikely to us that, as certain researchers have recently claimed, the eclipse of the Crucifixion was an eclipse of the Moon.

Prince Igor's eclipse

At the end of the eighteenth century, a masterpiece of Russian literature, *Slovo o polku Igoreve* (*The Tale of Igor's Troop*) came to light, recounting the expedition of Prince Igor Svjatoslavic against the Polovtsy in 1185. The tale inspired Aleksandr Borodin's *Prince Igor*.

The authenticity of this 'medieval text' caused great debate, and its discoverer, Prince A.I. Musin-Pushkin, a famous Russian antiquarian, was suspected of having written it. Chief skeptic was the French expert on matters Slav, André Mazon, whilst those who supported its authenticity as a twelfth-century epic included linguist Roman Jakobson and Soviet medievalist D.S. Liksatchev.

The Tale of Igor's Troop tells of an event only briefly mentioned in Russian chronicles of the Middle Ages: the doomed expedition against the Polovtsy in 1185, led by Igor Svjatoslavic, a descendant of Saint Vladimir and of Jaroslav the Wise, prince of the little city of Novgorod Severskiy (North Novgorod), on the Desna River to the north-east of Kiev. The Polovtsy, also known as the Koumans, were of Turkish origin, and had established themselves in the Don Basin and beside the Sea of Azov. The *Slovo*, about 3,000 words long, takes up the story of events in the chronicles, but differs from them in its literary richness, with its rhythmic phrasing, lyrical digressions and a certain epic feel.

At the outset, Igor and his forces are met by a bad omen: a total eclipse of the Sun. They decide to carry on regardless, joined by Igor's brother Vsevolod and his soldiers. More portents: wolves, eagles and foxes signal bloody defeat; the river is troubled; and dust obscures the sky.

After initial success, the Russians are threatened with encirclement. In a vague recital, peppered with invocations and oaths, the author recalls heroes, pitched battles of olden days, and ancient invasions by the peoples of the steppe. In spite of Vsevolod's bravery, Igor is beaten, and is taken prisoner with his brother and his son Vladimir.

In Kiev, supreme Prince Svjatoslav senses the catastrophe in a dream, and his boyars soon confirm it. Svjatoslav is furious at the two young princes' temerity: his own cousins, throwing themselves into a *démarche* doomed to failure. He despairs of the dissensions which split the princes of Russia, whom he names one after another. After this long digression, the author of the *Slovo* returns to Igor, who, helped by Vlur (in the chronicles, Lavor of the Polovtsy), manages to escape. Like a falcon, he flies off to his homeland, pursued by Koncak, Khan of the Polovtsy. Koncak finally relents: if he cannot catch the 'falcon' he plans to ally the Russians with his own people by marrying off his daughter to Igor's son, Vladimir.

Igor reaches Russian soil again, greeted by the chant of the virgins, and there is rejoicing in the towns. The *Slovo* ends here, but the chronicles tell us that in 1187, Vladimir returned home with Koncak's daughter, who bore him a child, and Igor blessed their union. (See frontispiece. Table 4.4 provides details of the eclipse.)

Table 4.4. The solar eclipse of 1 May 1185

First contact	13 h 37 m UT
Maximum	14 h 38 m UT
Last contact	15 h 37 m UT
Maximum phase	0.843
Local time for Novgorod Severskiy:	UT + 2 hours

Tintin's eclipse

In the volume of Tintin's well-known cartoon adventures entitled *Prisoners of the Sun*, his creator, Hergé, has Tintin, Captain Haddock and Professor Calculus saved from fiery execution by an eclipse of the Sun. Their funeral pyre, prepared by the Incas and due to be ignited by the focused rays of the Sun god, is set to dispatch our three heroes at a time which their captors have allowed them to choose. But the Incas' concession saves Tintin and his colleagues: Tintin has already discovered from a scrap of old newspaper that a total eclipse will be crossing Peru 18 days after their capture. We can try, like R. Moscovitch (Institut d'Astrophysique de Paris) to reconstruct the date of the eclipse and the location of Hergé's Temple of the Sun.

Tintin carried out some extremely complex calculations to be able to choose the exact day and hour of the sacrifice, and he must therefore have known the exact location of the Temple of the Sun: the patch of darkness cast by an eclipse of the Sun sweeps across several thousand kilometres, but the shadow is never more than a few hundred kilometres across. Conversely, if we know the date and time of the event, we can determine the location of the Temple of the Sun.

Hergé was known for his meticulous bibliographical research, and for the accuracy of the details found in his books. However, he provides no clue as to the year in which the exploits of his young reporter take place. To discover the location of the Temple, we must know the year in which the event is supposed to have occurred. The task is complicated by the fact that Hergé's adventures are not set in any precise timeframe; conversely, we are helped by the very rarity of an eclipse for any given place – in this case South America. Our search will range over a long period, of several decades. For reasons of historical verisimilitude, we can choose the year 1915 as a starting point. The final date for our research will be the date on which the adventure was published.

On 16 December 1943, *The Seven Crystal Balls* began to be serialised in *Le Soir*. Episodes appeared irregularly until 3 September 1944, at which time the liberation of Brussels intervened. Readers had to wait on tenterhooks for more than two years, until 26 September 1946, before they could find out what happened at the Temple of the Sun in the weekly *Tintin* magazine. Thus, we confine our research to the years between 1915 and 1950. The time of the eclipse is given in the story: 11 o'clock. However, since the Incas used sundials, we cannot be precise about the time, and must widen our research to allow for this.

Possible eclipses, seen from South America between 1915 and 1950, with details of countries concerned and coordinates of the locations of totality at 9, 10, 11 or 12 o'clock solar time, are listed in Table 4.5.

Careful calculations, based on simulations of the eclipses given in the table, suggest that the eclipse in question was that of 25 January 1944. This eclipse will

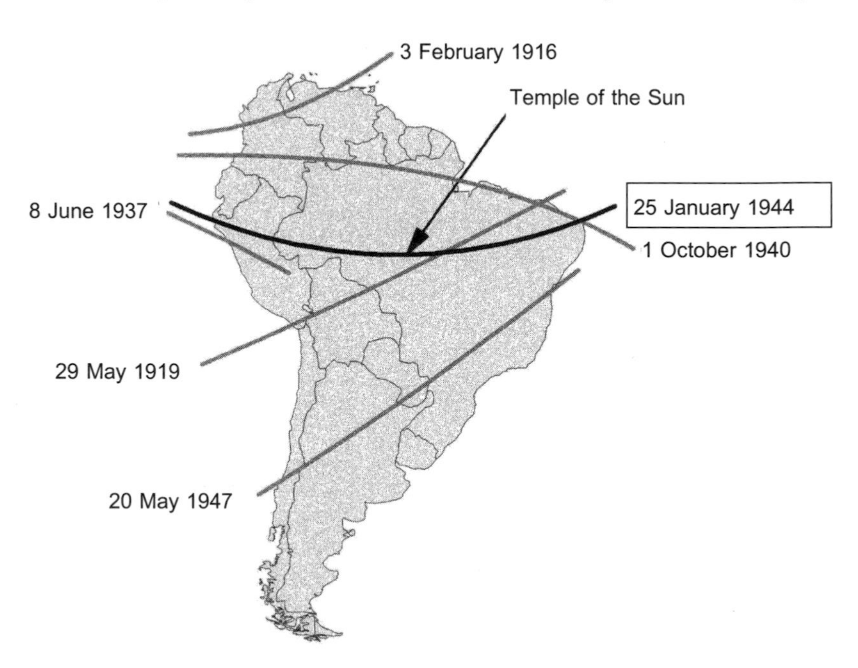

Fig. 4.6. Shown on this map of South America are the tracks of several eclipses which may have provided the inspiration for Hergé's eclipse in his cartoon story *Prisoners of the Sun*. The most probable is that of 25 January 1944, and the location of the Temple of the Sun is marked by an arrow. (P. Guillermier.)

locate the Temple of the Sun at longitude 58° W and latitude 9° S, but this is hardly a likely site, as the Temple would be about 700 km south of Manaus, in Brazil, and very far from Inca territory.

It seems likely that the author of these adventures was inspired by the eclipse of 1919, which was widely reported in the media at the time, and we must conclude that Hergé was probably describing an imaginary eclipse; the site of the Temple of the Sun will remain forever a mystery.

Table 4.5. Solar eclipses over South America

Date	Countries	Coordinates of site of totality
3 February 1916	Colombia, Venezuela	12 h, Long. 67° W, Lat. 11° N (Caracas)
29 May 1919	Peru, Chile, Bolivia, Brazil	9 h, Long. 40° W, Lat. 4° S (between Recife and Belem)
8 June 1937	Peru	No eclipse for hours selected
1 October 1940	Bolivia, Venezuela, Brazil	10 h, Long. 30° W, Lat. 10° S (Atlantic Ocean)
25 January 1944	Peru, Brazil	11 h, Long. 58° W, Lat. 9° S (south of Manaus)
20 May 1947	Chile, Argentina, Paraguay, Brazil	No eclipse for hours selected

Nat Turner's eclipse

Nat Turner, a black slave in nineteenth-century Virginia, fomented a rebellion among his people which has gone down in history alongside those led by Gabriel Presser in 1800 and Denmark Vesey in 1822. Turner was born in 1800 in the USA, on a Virginia smallholding. His father disappeared when he was only a young boy; and his mother, who had come from Africa, instilled into him a fierce hatred of slavery. He learned to read alongside his master's children, and was passionate about religious education, which was to lay the foundations of many of his later deeds. Working as a carpenter on the plantations, he was sold twice – in 1820 and around 1830 – though his hope was to be freed. He married a young slave from another plantation. In 1820, he ran away and hid in the swamps for a month, but returned to the plantation, called, as he claimed, by divine inspiration. He then began spreading the ideas championed by Senator Rufus King of New York, and the Quakers, declaring slavery to be against the laws of God and nature. Turner dedicated his life to the goal of emancipation.

In 1825 he began to preach, and, endowed with a certain magnetism, identified the lot of the slaves with the plight of the Israelites of the Old Testament. He spoke of the justice and vengeance of God being visited upon the oppressors. The attitude of the whites caused him to think that only armed force could save the black people, and, claiming divine guidance, in May 1828 he decided to launch a rebellion. As a noted preacher of the region, he gathered around him a small group of about 20 slaves devoted to his cause, determined not to suffer the same kind of failure as had Denmark Vesey.

Fig. 4.7. In 1831 the religious leader Nat Turner led a slave uprising in Virginia. He told his followers of the vision he had had, during the 1831 annular solar eclipse, of a black angel overcoming a white angel, signalling to him that the time for rebellion had arrived. (Schomburg Center for Research in Black Culture.)

To Turner, the eclipse of the Sun on 12 February 1831 came as a divine sign, ordering him to act. On 21 August he set off with seven other slaves on a bloody journey, attacking any white person encountered on the way. In two days and nights, 51 whites were murdered, and about 75 blacks joined in his march on Jerusalem, the main town of Southampton County.

The Virginia state militia were mobilised, and several thousand men hunted down the rebels, who were dispersed, arrested or killed. Turner, having gone into hiding for six weeks, was finally captured. At his trial, he pleaded not guilty, but he was hanged. A wave of fear then swept through the white population of Virginia, and slaves were arbitrarily killed. The state of Virginia passed implacable laws further restricting the slaves, and they were forbidden to preach or to meet together. These laws were in force until the Civil War. The name of Nat Turner was long venerated in the South, and the mention of Jerusalem in the songs of the region recalled the Turner rebellion rather than the biblical city. (Table 4.6 provides details of this central annular eclipse.)

Table 4.6. The solar eclipse of 12 February 1831

First contact	16 h 19 m UT
Maximum	17 h 53 m UT
Last contact	19 h 24 m UT
Maximum phase	0.945
Local time for Jerusalem, Maryland:	UT – 5 hours.

Fig. 4.8. The solar eclipse of 22 May 1724. (Paris Observatory.)

Eclipses and early discoveries

The study of the solar corona, recognised as an entity for more than a century now, has a history rich in discoveries and full of anecdotes.

Aspects of the story of these observations have been recounted in works on the corona and the Sun's corpuscular emissions by, for example, I.S. Shklovskii, D.F. Billings, E.N. Parker, S.K. Vsekhsvjatsky and J.C. Brandt (see Bibliography). More recent books by J. Zirker and by L. Golub and J. Pasachoff continue the tradition. We can select only a few of the best known episodes here, after sifting through original documents at the Paris Observatory.

A story dating back to the eclipse of 22 May 1724 tells of a marquis arriving at the Paris Observatory a few minutes after the event, in the company of some noble ladies. He told them: 'Let us enter nonetheless, Mesdames; Monsieur Cassini is a great friend of mine, and will be only too pleased to repeat the eclipse for you.' A contemporary painting shows Parisians crowding around the Observatory, observing the eclipse with the most unlikely apparatus: sieves, funnels, musical instruments ... and, of course, the traditional smoked glass and buckets of water. A planet – probably Venus – shines in the darkened sky.

One of the first exhaustive studies to examine observations of total eclipses was carried out by F. Arago (*Annuaire du Bureau des Longitudes*, 1842). Arago quotes a description, attributed to Plutarch, of the eclipse of 98 AD: 'The Moon allows part of the Sun to spill out all around her, and darkness is decreased ...', which shows that such phenomena are nothing new; similar clues are found among the remains of pre-Columbian civilizations, especially the Aztecs. Apparently, the Ancient Greeks adjusted their calendars by calculating precise dates from their observations. But, apart from these purely mechanical pursuits, the study of the phenomenon of eclipses has led, since the late nineteenth century, to the identification and analysis of the Sun's environment. Solar physicists 'cover' these events, as do geophysicists, interested in the disturbances caused to the Earth's upper atmosphere during eclipses. It should be remembered that, as late as the mid-nineteenth century, astronomers had been unable to establish any link between the Sun and the aureole surrounding it during a total eclipse.

In Western tradition, the history of the solar corona is a curious one. From the sixteenth century until the late eighteenth century, the corona was deemed to be merely a halo or coma caused by Earthly fumes, or the Moon's atmosphere. Kepler's observations of 1567 led him to think that this 'coma' was the result of sunrays refracted in the air of the Moon. Later, Edmond Halley, observing a total eclipse from Greenwich on 3 May 1715, and 'recognizing' that the aureole was 'exactly' centred on the Moon, interpreted it as the lunar atmosphere. The great man was let down by his micrometer, and too high a magnification. Even Arago (*op. cit.*) was proposing in 1842 to call the phenomenon the 'lunar corona'! This understanding of the nature of the Sun's corona highlights the great efforts of certain pioneers (among them, the same Arago who indeed performed the first measurement of the polarisation of the corona in 1842) who showed remarkable insight into the problem, mostly through observations of the zodiacal light. It would not be too far from the truth to state that the idea of an almost continuous corpuscular stream flowing from the Sun originated in the late eighteenth century. It was Cassini who, having identified the nature of the zodiacal light through parallax measurements in 1683, and having determined from numerous observations that its brightness fluctuates, deduced that: 'our [zodiacal] light would seem to have the same vicissitudes as have the sunspots, which were very frequent in recent years; which appears to support the idea that, in some way ... this [zodiacal] light may be part of one flux with the spots and faculae of the Sun.'

It should be noted that, just as the origin of the zodiacal light was once unknown, so its fluctuations are nowadays not definitely established. Descartes imagined that it owed its existence to emanations from the Sun: '... a kind of effervescence or purging of the baser parts.' An alternative hypothesis was proposed by Newton, who thought that it was 'a mass of heterogeneous particles spread in the ether, drawing in upon themselves and falling towards the Sun.' It is now known that fluctuations observed from the ground are in fact of geophysical origin (auroral phenomena known as the 'false zodiacal light').

The greatest progress, however, was made by M. De Mairan, who in 1731 established a link between fluctuations in the brightness of the zodiacal light and the frequency of auroral displays. His ideas were published in a remarkable memoir in 1733. He offered a very simple explanation for the circumsolar nebulosity, affirming that 'the zodiacal light is no more than the atmosphere of the Sun, a fluid or very rare and tenuous matter, self-luminous, or lit merely by the rays of the Sun, and encircling the globe of the daystar, being of greater abundance and extent around its equator than elsewhere.' De Mairan also studied the links between the occurrence of sunspots and the frequency of aurorae, and suggested a mechanism describing the action of the zodiacal light on the Earth's atmosphere, thought to be responsible for the phenomenon. The memoir is supported by many references to observations, some of them systematic, which elevate the work above the merely prophetic, in spite of its speculative statements on the nature of the zodiacal light. De Mairan therefore had a first inkling of the phenomenon of the Sun's mass loss, or what is now called the solar wind. The gaseous nature of the coronal phenomenon remained a matter of debate, with 'metallic vapours' sometimes being evoked. Elsewhere, scientists claimed that the coronal halo seen during an eclipse was due to high-altitude smoke from fires, lit by superstitious primitives trying to frighten away evil spirits.

On 15 May 1836 the Moon's shadow moved across southern Scotland. English astronomer Francis Baily saw this annular eclipse, and noticed irregularities at the Moon's limb as the Moon covered the Sun. He described seeing the light of the Sun resembling a string of beads, and explained the phenomenon thus: at the Moon's edge, light from the photosphere is blocked by high mountains, and shines through the valleys. The term *Baily's Beads*, which arose from the description, remains in use today.

Since Baily's eclipse, many astronomers have positioned themselves within the Moon's shadow to observe the phenomenon. They have trekked to remote places, hauling tonnes of equipment, to observe for those few fleeting moments, and to uncover more of the secrets of the Sun.

The first observation of shadow bands – alternating light and dark areas due to atmospheric interferences – is ascribed to H. Goldschmidt.

The first attempt at eclipse photography, on 8 July 1842, was by an Austrian, Majocci, who succeeded in obtaining images of the partial stages, but failed to capture totality. The first photograph of an eclipse to succeed in showing the corona and prominences was taken by Berkowski on 28 July 1851, with a 6¼-inch heliograph at Königsberg. The exposure time was 24 seconds.

A mystery was solved during the eclipse of 18 July 1860, when, at two sites 500 km apart, the Englishman Warren De La Rue and the Italian Angelo Secchi used the relatively new technique of photography to record prominences. At that time, many astronomers still believed that prominences were lunar in nature. However, photographs taken at the two sites showed no parallax shift for the prominences, and thus it was demonstrated that they could not be part of the Moon, but were in fact elements of the Sun.

From 1842 onwards, total eclipses were the object of intense study, and the true nature of the Sun's surroundings was revealed. Without a doubt, the definitive instrument of this revelation was the prism spectroscope. The realisation of the solar nature of the corona was probably due to several observers reporting the same prominence at the same eclipse, but at different times and places. Methods were many and varied. For example, Jules Janssen worked with a helioscope with a 50-cm objective at f/3.2. Later, he built a 1-metre telescope, but this was never used for eclipses.

During the eclipse of 1868, and more especially that of 1871 in northern India, Janssen carried out observations with a view to determining the nature of the solar corona. These are without doubt the best visual observations ever done: 'The

Fig. 4.9. Equipment for observing the eclipse of the Sun of 18 August 1868. The French expedition of Stephan, Rayet and Tisserand was stationed in the Kingdom of Siam, on the Malacca peninsula, with the large 40-cm Foucault telescope, a 20-cm telescope, and a number of spectroscopes. Each member of the expedition became famous in his own field: Stephan went on to observe the deep sky, and more especially the quintet of galaxies named after him; Rayet became known for his studies of the emission lines of certain stars; and Tisserand carried out remarkable work on celestial mechanics, particularly concerning comets. (From a photograph by M.G. Rayet, Paris Observatory Library/Astronomie-SAF; document selected by S. Dumont in 1981.)

spectrum of the outer regions of the corona shows immediately the very remarkable green line already mentioned. In the middle part of the corona, from three to six arc minutes, the dark D line is perceived ... This observation confirms the presence in the corona of reflected solar light, but one senses that this light is subsumed in an abundant and bright emission of another origin ... the slit was placed in such a way as to intercept a part of the Moon, a prominence, and the whole height of the corona. The Moon's spectrum appears excessively pale, and seems due mostly to the brightness of the atmosphere ... the prominence offers a very rich spectrum of great intensity ... the main point here is that the principal lines of the prominence are continued throughout the corona, demonstrating decisively the existence of neutral hydrogen in the corona. The green line, so evident in the coronal spectrum, seems to disappear in that of the prominence.'

Janssen's observation of the eclipse of 6 May 1883, which crossed the Pacific Ocean, confirmed his previous deductions. He obtained a complete Fraunhofer spectrum of about 100 lines: 'The phenomenon was evident for the brightest parts of the corona, but not for the very base, where the spectrum appeared continuous. The phenomenon was not of the same intensity at places equidistant from the Sun's limb.' These extracts from Jules Janssen's work show that he had, as early as 1871,

Fig. 4.10. The great 40-foot (12-metre) Lick Observatory eclipse camera, set up near Jeur in India for the eclipse of 22 January 1898. This giant instrument – named 'Jumbo' by its creator, J.M. Schaeberle – was transported to every continent, with the exception of Antarctica, for every eclipse from 1893 to 1931. It could not be guided in the normal way (although it had a clock-driven plate-holder), and was aimed at the pre-calculated position of the eclipse. It could provide a dozen photographs per eclipse on 56 × 36-cm glass plates. (High Altitude Observatory collection.)

identified the F (Fraunhofer) component in the solar corona. Before this, he had identified prominences during a solar eclipse; and he also devised a method (independently of, but at the same time as, J. Norman Lockyer), from the eclipse site itself, of observing them when the Sun was not eclipsed ('I shall be seeing these lines again . . .'). Interestingly, it should be noted that some over-skeptical observers of the period, in spite of the massive equipment at their disposal – such as the Lick Observatory 40-foot (12-metre) instrument – were still far from reaching this point in their research. In 1883 Professor Hastings wrote: 'Has the corona any objective existence? Is it not merely an optical illusion produced by diffraction?'; while others, such as Young, had made the first spectroscopic observations in 1869 and had discovered the new line attributed to a new element, given the name 'coronium'.

Photography was eventually to produce a harvest: Huggins' description of the corona in 1885 is in accordance with modern findings: 'The corona is a structure of great complexity, and even more so in its three dimensions.'

But the most remarkable deduction is, in our opinion, that of A. Schuster. In 1891 – more than 20 years before the first measurement of solar magnetic fields using the Zeeman effect – he wrote: 'The form of the corona at the time of minimal sunspot activity is, in fact, very similar to that which would be obtained if the Sun were a

"Waiting for the Eclipse"

Fig. 4.11. An illustration from *Nature*, 1 February 1872, showing astrophysicists of the period preparing their instruments. (*Nature*.)

magnet, discharging negative electricity at its polar regions.' The existence of a general magnetic field for the Sun, dictating the shape of the corona at the time of minimum activity, had been postulated in 1889 by Bigelow, a physicist at the Smithsonian Institute at Harvard.

Finally, we offer two anecdotes which demonstrate the tenacity of astronomers in their desire to observe eclipses. Jules Janssen was in Paris during the siege of the city by the Prussians, and escaped in a balloon in order to observe the eclipse of 22 December 1870. Having arrived safe and sound in Oran, North Africa, he was prevented by clouds from observing the event from the ground. In an effort to counter the vagaries of the weather, D.I. Mendeleev – the well-known physicist who devised the periodic table of the elements – went up in a balloon to an altitude of 3,500 m to observe the Russian eclipse of 19 August 1887.

Twentieth-century eclipses

For most of the first half of the twentieth century, progress in coronal physics relied essentially on spectrophotometric work, photography, and the systematic study of Sun–Earth interactions, in which area most new data were obtained. Geomagnetic disturbances like aurorae were interpreted as manifestations of corpuscular emissions, and many studies were dedicated to these emissions. Between 1908 and 1913, K. Birkeland attributed them to electrons alone, suggesting that 'corpuscular rays' from the Sun were the cause of geomagnetic disturbances and the aurora borealis. In 1911, after extensive investigations, C. Störmer offered an explanation of the morphology of the corona at solar minimum: the calculated paths of electrons ejected from the solar surface in a dipole field localised at the centre of the Sun represented perfectly what was observed in the corona at the minimum of activity. Although sophisticated, this exercise was somewhat academic in that it is not borne out by physics; in fact, the gyromagnetic radii of coronal electrons have much smaller values than Störmer supposed. However, the idea of the corona as a gas of electrons was beginning, at that time, to approach reality, and opened the way to the explanation of the amounts of polarisation observed in the white-light corona. Moreover, the magnetic origin of coronal structures was implicit in Störmer's calculations. In the same year, A. Schuster showed that the corpuscular 'beams' must be electrically neutral – containing electrons and protons. In 1919, Lindemann suggested a new theory taking this objection into account, but this view of the coronal gas did not gain support at the time. Later, in 1931, Chapman and Ferraro traced out the basics of the theory of the magnetosphere, and then H. Alfvén lent the weight of his opinions to the theory of cosmic plasmas.

Henceforth, it was accepted that the Sun emits particles, and the large coronal streamers were mostly identified with these emissions. An example of this is the little-known model proposed by S. Rosseland in 1933. He applied the equation of hydrodynamic continuity to the zone at the base of the corona, and 'found' the continuous corpuscular emission of the Sun. In the introduction to his paper *On the Theory of the Chromosphere and the Corona*, Rosseland sums up the situation at the time, which seems not to have changed very much to this day: 'Chromosphere, corona and prominences ... form a complex of dynamic phenomena, the theory of

Fig. 4.12. A small accessory for the eclipse of 17 April 1912. On this filter, which allowed safe observation of the Sun, there is also a reference to the eclipse of 11 August 1999. (Kindly donated by B. Raband, P. and M. Curie University, Paris.)

which must be based upon considerations of the expansive motion of matter moving away from the Sun, in a more or less radial direction.' Rosseland and many other researchers of the solar corona tried to show that large coronal streamers are the source of 'continuous' corpuscular emissions from the Sun. More recent theories have tended to contradict this. Between 1958 and 1962, E. Parker led the way in piecing together the theory of the solar wind, showing that hydrodynamic solutions could be found to more general equations describing the equilibrium of gas at coronal temperatures, with account being taken, on the one hand, of its expansion and dynamic pressure, and on the other, of the equation of conservation of mass.

However, this highly successful theory, which describes from simple assumptions both the supersonic radial expansion of coronal gases and the expansion of magnetic field lines caught up in this flow, is not valid *a priori* for the non-collisional medium to which it is applied. It does not explain how or why the corona is heated to millions of degrees, which is certainly the true reason for the expansion. Solar activity and magnetism are responsible for corpuscular emissions, as many astrophysicists suspected in the 1950s. We have, for example, the forward-looking work of J.-C. Pecker and W.O. Roberts, who postulated the existence of particular regions of the solar disk, identified with the M regions (magnetic, or unipolar) which are responsible for recurrent magnetic disturbances. The idea of coronal holes was first advanced by them in a paper published in a geophysical journal.

The existence of the solar wind was definitively established by *in situ* measurements made by early Soviet and American space probes, whilst, much earlier, Bierman had been the first to calculate the velocity of the solar wind to correct values (> 300 km s^{-1}), with a theory explaining the deviation of the ion tails of comets approaching the Sun.

In the next section we touch upon some memorable eclipses of modern times, from which astrophysicists have reaped a rich harvest.

Einstein's eclipse

We begin with a famous episode, even though it has no direct link with solar physics. Its result might be considered to be only a by-product of eclipse observation, but it led to considerable excitement during the early part of this century.

In 1905, Albert Einstein, then 26 years old, published several papers in *Annalen der Physik*, on the subjects of Brownian motion, the photoelectric effect, and special relativity. This work transformed the obscure Berne patent-office clerk into a world-famous scientist. During the next ten years, Einstein described, in his theory of general relativity, the distortion of space-time in a strong gravity field. This theory appeared in its final form between 1913 and 1916. It predicted that the curvature of space, near massive objects like the Sun, would bend light passing in its vicinity. A star, observed near the Sun's limb in favourable observing conditions – such as during a total eclipse – should appear displaced by 1.75 arcsec from its expected position. A German expedition was organised to observe the total eclipse of the Sun of 21 August 1914, in order to verify this prediction. However, the German scientists were arrested on Russian territory and interned for a month before being released, and no photographs of the eclipse were obtained.

When, after World War I, the great British astronomer Arthur Stanley Eddington heard about Einstein's prediction in the context of general relativity, he immediately saw the importance of confirming this discovery, and mounted an expedition to observe the eclipse of 29 May 1919. On that day, light from the Hyades star cluster would pass close to the Sun's disk, on its journey to the observers. In order to increase his chances of success, Eddington chose two observing sites. One group of astronomers, led by Andrew Crommelin and Charles Davidson, would observe from Sobral in Brazil, and the other, including Eddington himself and Edwin Cottingham, would be on the island of Principe, off West Africa. On Principe, the day of the eclipse began with heavy rain, which stopped early in the afternoon; the Sun appeared a few moments after first contact. Undaunted, Eddington seized his chance and took 16 photographs through broken cloud. About six months later, he repeated his photographs at night, for the purposes of comparison. The advanced state of photography at this time is well illustrated by the fact that the images of the solar corona clearly showed fairly bright field stars ($m_V < 9$) during totality.

The results wholly verified Einstein's theory. They were announced on 6 November 1919 at a meeting of the Royal Astronomical Society. The photographs show a displacement in relation to the Sun's edge of between 1.65″ and 1.98″.

Fig. 4.13. This press cutting shows the Eddington expedition's observing site at Sobral, in Brazil, at the time of the eclipse of 29 May 1919. The diagram explains the deviation of light rays from a star as a result of the curvature of space near the Sun. At a joint meeting of the Royal Astronomical Society and the Royal Society in London, Arthur Stanley Eddington announced his results, confirming that Einstein's theory superseded that of Newton, whose portrait looked upon the meeting. (*Illustrated London News*.)

All over the world the press reported the findings, with some amplification, and dubbed Einstein a genius. There is no doubt that Einstein would not have received such legendary status without the media attention lavished upon the success of his prediction. Nowadays, the phenomenon of the gravitational 'mirage' goes far beyond the vision of its gifted interpreter: it is now recognised on a scale stretching across the cosmos. But that is another story.

Eclipses of the 1930s

The art of photographing the corona of the eclipsed Sun reached great heights during the 1930s, though the achievements of the years 1960–1970, and those of the later era of colour images, were still to come. What was remarkable during this inter-war period was the size of the instruments and of the photographic plates used.

Most noteworthy were several cameras, equipped with objectives corrected for chromatic aberration, with diameters of the order of 20 cm, and focal lengths of about 20 metres, often aimed at the Sun from the tops of towers. The plates in the focal plane, 40 or 50 cm across, were moved on an inclined table, to compensate for diurnal motion during the exposure. The logistical back-up employed by eclipse expeditions during this period was impressive, with an expeditionary force of military personnel, usually naval and airborne; details of secondary importance perhaps, although their presence was justified if observations took place in unstable countries, or even in deserts. The stories of some of these expeditions are so eventful that to tell them would require far more than the space allotted for the whole of this historical section! Among the most extensive are the American expeditions, with financial backing from the National Geographic Society, which publishes accounts of the eclipses in its famous magazine, together with results including, of course, images of the corona. Total eclipses of the Sun have become public property worldwide, and our interest is fuelled by the amazing and beautiful images produced by science. This has grown into a tradition, more active as the years go by. Let us not forget the presence of at least one graphic artist on major expeditions, with the task of immortalising in a few minutes the fugitive vision of the corona during totality.

Since the beginning of the century, astronomers have systematically taken up the challenge of photographing eclipses, since drawing the phenomenon over such a brief interval is problematical. One vital outcome of this new technique would be to affirm the reality of the details reported by the best visual observers: the chromospheric fringe, prominences, arches, giant streamers, plumes and so on. It soon became apparent that the enormous brightness gradients, or radial intensities, across the corona meant that not all details could be captured on one plate. To counteract the effects of diffusion from overexposed areas on plates, high magnifications with very long focal lengths were used, as well as plates with a very low contrast factor (gamma) of the plate's response curve. But capturing the image with one exposure remained difficult. So the compiling of a faithful and highly detailed image through drawing was necessary, based on examination of the plates by several specialist artists. By the late 1930s, this particular art form had attained a level of near-perfection in Russia and America.

In Europe, drawing was commonly used in the late nineteenth century to depict solar coronae; the work of Father Secchi is a case in point. As an aid to the artists, images using an enlarger with a parafocal mask were sometimes created, working from originals taken with different exposure times. Happily, the art of solar drawing flourished, supported by photography, allowing the structure of the corona to be analysed in conjunction with activity on the solar disc. This led to an appreciation of the importance of the solar magnetic field in the corona, and to the establishment of a basis for eventual numerical modelling; but more objective data were still lacking – data which only photometry and polarimetry could supply. Nevertheless, the plates could supply a good amount of detail, and unveiled the links between the chromosphere, the prominences and some details of the inner corona, and of the mid-corona: for example, occasional darker zones were detected, and interpreted as cavities in the corona, around prominences; or, on a smaller scale, the apparent continuity between a large chromospheric spicule and the fine radial streamers composing plume-like structures. This kind of observation is still prized even today, as the interface between, on the one hand, chromosphere and prominences, and on the other, the inner corona, is still incompletely understood. This region – highly dynamic, even turbulent, and of widely varying temperatures – is the seat of small-scale areas of particle acceleration, and of magnetic 'traps' (topological singularities). During the 1930s, significant technological progress was made, not least in the field of research into colour emulsions, somctimes with curious results such as the acquiring of blue or green images of undoubtedly red prominences! The problem of overexposure remained critical, and it was during this decade that the first attempts were made to incorporate radial-gradient neutral-density filters of the mechanical type, later reintroduced by M. Laffineur for the 1961 eclipse. A similar device had been tried out by Professor Burckhalter of the Chabot Observatory in 1900, and then by Professor R.W. Marriott of Swarthmore College in the USA, for the eclipse of 14 January 1926, with excellent results. It was the Swarthmore group who constructed a massive 22-cm tower instrument, of 17 m focal length, in order to obtain plates of coronal images.

A wide-ranging article on the eclipses of 1936 and 1937, by Irvine C. Gardner of the National Bureau of Standards in Washington, appeared in the special eclipse number of the *National Geographic*, No.1, 1939, and describes the various instruments used, included a 'rotating sector disc', successfully used at the eclipse of 8 June 1937. At this time, the system was employed at the focus of a camera of 5 m focal length, together with an eclipse coelostat. This was a similar array to the multiple instruments used by Soviet teams, who had brought back memorable results from the preceding eclipse of 19 June 1936. A large American expedition, led by well-known astrophysicist D.H. Menzel, travelled to the USSR for the 1936 eclipse, which crossed the whole width of Soviet territory and therefore offered continual views of the corona over some two hours. Eclipse experiments by Taffara and Horn d'Arturo, of Bologna University, in 1923 and more especially in 1926, had led to the first comparison of photographs taken at intervals of nearly 2.5 hours, and these had provided some insight into proper motions observed in the corona. Values were established, from 0.6 km s^{-1} in the inner corona to 150 km s^{-1} for radial motions. Nevertheless, the dynamic nature of the structures remained largely unexplained,

Fig. 4.14. The eclipse of 14 January 1926. The solar corona, drawn from photographic plates of various exposures by Emma T.R. Williams. (J.A. Miller and R.W. Marriott, Swarthmore College.)

and the rotation of structures with the Sun could still not be measured, or even understood. Remember that, at the time, the corona remained very much a mystery. Its gas was still thought to be at temperatures comparable to that of the surface, given the decidedly white colour of the corona. The most common hypothesis was that the atmosphere of the corona consisted of a gas made solely of the lightest known particles – electrons – because the light was polarised and the radial density gradient was manifestly weaker than that of the upper photospheric layers, where the scale in altitude is of the order of only 100 km. Thus, the charged corona would have a very high electrical potential, which caused some problems for the theorists! The reasons for the existence of coronal structure were still unclear, even though observation of dynamic phenomena in prominences had already shown trajectories incompatible with mere ballistic flight. The influence of other forces was suspected. It was in this context that renascent Soviet astronomy embarked upon an ambitious programme for the eclipse of 1936, extended later to that of 1941, directed by several astrophysicists who were to become famous – if they were not liquidated! The principal director for the whole programme was B.P. Gerassimovitch, director of the Pulkowa Observatory, who, after vicious denunciations were made against him, disappeared a few years later, swept away by Stalin's purges.

The scientific results of this very extensive observing programme were nevertheless published in several collections, some in English, and even during the war. The book by E. Bugoslavskaya on the structure of the solar corona is full of references to the programme, and results were comparable to the eclipse of 1941.

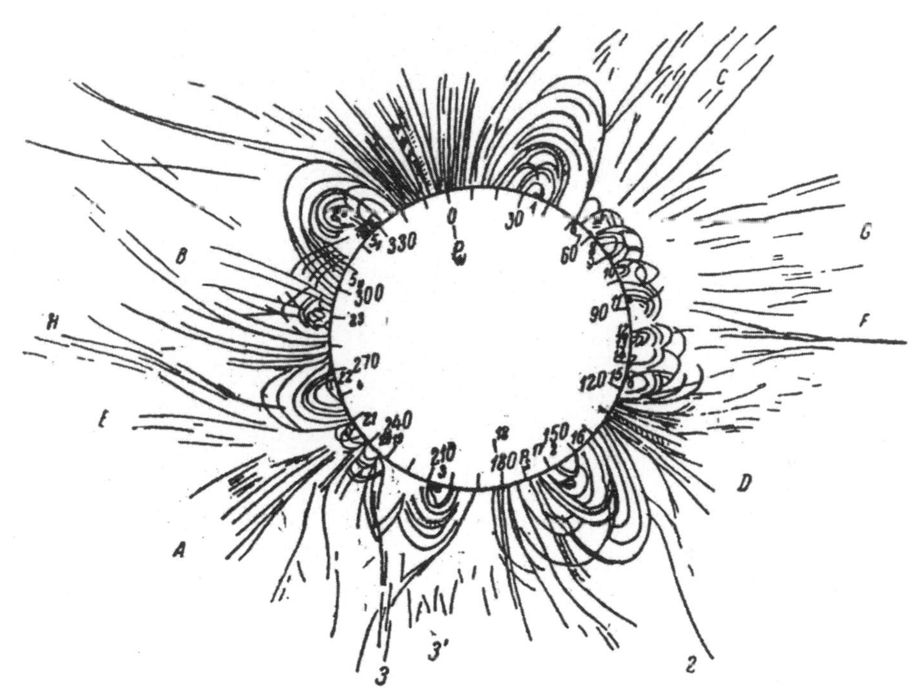

Fig. 4.15. Drawing of coronal structures of the eclipse of 19 June 1936, from several plates taken by the Soviet teams. The instruments included a coelostat and a 15-cm doublet of 6 m focal length. (S. Vsekhsvjatsky, Kiev, and E. Bugoslavskaya, Moscow. From eclipse observations, 1936.)

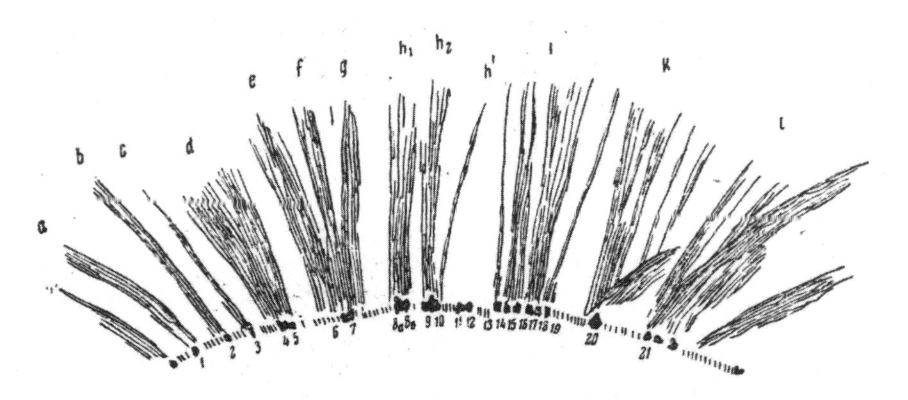

Fig. 4.16. Drawing of chromospheric structures and coronal extensions at the north pole of the corona, during the eclipse of 21 September 1941. (E Bugoslavskaya, Moscow University.)

Large numbers of motions of coronal structural 'details' could be measured, and 'speeds' of between 0 and 10 km s^{-1} were derived for details of the inner corona. The result which had the greatest impact was that the inner corona exhibited mean velocities compatible with a synchronous rotation of the entire corona and the disk. This measurement was of great importance, proving beyond doubt the existence of a force acting upon the corona to make it rotate in a quasi-rigid manner; and this force is obviously magnetic.

To further enhance the picture of the eclipses of this decade, we should also mention the French teams, who had mixed fortunes with their observations. For the 1936 eclipse just mentioned, a large expedition, planning various experiments, went to Soviet Kazakhstan. They brought back no results, however, having been beneath a thick cloud layer throughout totality. Several famous (or soon to be famous) astronomers accompanied this expedition, among them R. Dufay (director of Lyon Observatory), M. Gauzit (professor at Montpellier), D. Barbier (assistant astronomer at Marseille), Madame Chalonge (assistant astronomer at the Paris Observatory), and Dauvillier (Collège de France). Bernard Lyot had also prepared an experiment, but did not make the journey, as he was working on a new coronagraph for the Pic-du-Midi Observatory.

In spite of the ignominious lack of results, this eclipse did prove beneficial, in that it brought together D. Barbier and D. Chalonge, who became founder members, with H. Mineur, of the Institut d'Astrophysique de Paris at the time of the creation of the Centre National de la Recherche Scientifique (CNRS).

One more expedition to be mentioned is that of the Bureau des Longitudes, to the eclipse of 9 May 1929. Their site was the island of Poulo Condore (Indochina), and the team included famous astronomers such as A. Danjon (later director of the Paris

Fig. 4.17. This photograph shows R. Dufay (centre), M. Gauzit (left), Madame Chalonge (mid-distance) and D. Barbier (at back), with their telescope for the eclipse of 19 June 1936, at Kustanaj, Kazakhstan. (IAP/CNRS.)

Fig. 4.18. D. Barbier prepares his instrument at Kustanaj, under the curious gaze of some young Kazakhs, before the Kazakhstan eclipse of 19 June 1936. (IAP/CNRS.)

Observatory) and A. Lallemand (later director of the Institut d'Astrophysique de Paris).

The aim of eclipse work was not only to achieve photographic results: spectroscopic analysis was also on the agenda, and a specialist in this field was S.A. Mitchell, author of the first complete book on solar eclipses, published in 1923 and revised and reissued several times until 1951. The famous 'flash' spectra which were obtained revealed the chromosphere in a new light, and gave rise to the term 'reversing layer'. Several decades passed before scientists understood the complex thermodynamic phenomena producing emission lines in the upper layers, showing that the temperature was increasing with altitude in the solar atmosphere, in association with an ever more turbulent and anisotropic velocity field. The most modern telescopes now have the capacity to investigate these problems by direct observation even when there is no eclipse.

Bernard Lyot's eclipse

In 1945, towards the end of World War II, a total eclipse was due, with good chances of visibility, especially from Sweden, where K. Lundmark had already established a reputation. At the urging of D. Chalonge, France decided to send a team only two months before the event, with a view also to re-forging international links. Chief among the members of this expedition was Bernard Lyot, whose work with the coronagraph, the polarimeter and the Lyot birefringent filter had gained him an international and enviable reputation. Lyot took a polarimetric telescope with him

Fig. 4.19. Bernard Lyot stands by his polarimetric telescope, set up near Brattas in Sweden, before the eclipse of 9 July 1945. (IAP/CNRS.)

to Sweden. The French expedition was blessed with clear skies on the day, but their rapidly-prepared experiments yielded but few results – although this is understandable considering the difficulties of the period.

In 1947, before Commission 13 (solar eclipses) of the International Astronomical Union, Lyot, then France's youngest *académicien*, and already famous, described the merits of his spectrograph, destined for eclipses: 'Spectra obtained will provide the opportunity to study the distribution of lines detected only during eclipses, and results can be matched with theory.'

Lyot's intention was to design and use a curved-slit spectrograph to study the composition of the solar corona during a total eclipse of the Sun across wavelengths from red to ultraviolet – difficult to observe with his existing coronagraph, in use at the Pic-du-Midi Observatory for observations of the strongest lines.

In 1950 Professor Madwar, of the Egyptian Royal Institute, invited Lyot, with the promise of considerable material support, to mount a joint expedition to observe the eclipse of 25 February 1952 in Khartoum, capital of the Sudan, then an Anglo-Egyptian territory. In just a few months, Lyot had to construct two curved-slit spectrographs – one for visible light and one for ultraviolet. This was, however, not the only tool of the French astronomers: the fast-developing science of radio astronomy was represented by the presence of a 6-m radio telescope, to collect data at wavelengths of 55 cm and 117 cm.

As well as these two experiments, there were to be many others involving heliography, low-resolution spectrography, photography and monochromatic polarimetry. The French contingent included young astronomers who were later to become famous: Raymond Michard (past president of Paris Observatory), Jean-

Fig. 4.20. In this photograph, taken at Khartoum in 1952, are Bernard Lyot (front left), Professor Madwar (front right), and Khairy Aly (extreme right). (J. Leclerc, from the collection of G. Lyot.)

Claude Pecker (former director of the Institut d'Astrophysique de Paris, and today member of the Academy of Sciences), Audouin Dollfus (now famous for his planetary research), and Marguerite d'Azambuja. With the help of radio expert André Cassagnol (from the Steinberg team), electronic engineer Marius Laffineur took charge of the party from the Bureau des Longitudes, using a portable radio telescope of his own design. In Europe, Ivan Atanasijevic had helped with the preparation of the mission, and would be instrumental in the interpretation of its resultant data.

The ship *Champollion* berthed at Alexandria, with Lyot and his instruments on board, on Christmas Day, 1951. Egypt was living through troubled times, with King Farouk's government organising resistance with the aim of driving out British troops. Clashes and riots were commonplace, and on 26 January 1952 martial law was declared and Egyptian personnel went on strike. Undaunted, Lyot set up his spectrographs to the west of Khartoum, together with those of Egypt's Helwan Observatory. The French team set up their headquarters at the Fort Stanley site, east of the town, where there were other foreign parties. Dauvillier set up not far from Lyot's site, to take photographs of the white-light corona. On 22 February 1952, three days before the eclipse was due, a desert sandstorm blew up and raged for 24 hours. After the passing of the storm, as the sky reverted from a reddish hue to blue, it was found that dust had penetrated everywhere, in spite of the tents erected to protect the instruments. Every instrument had to be dismantled and cleaned.

Finally, on 25 February 1952, at 11 h 8 m, the Moon's shadow raced across the dunes. Totality! Lyot wrote enthusiastically: 'The corona has a large tuft to the west,

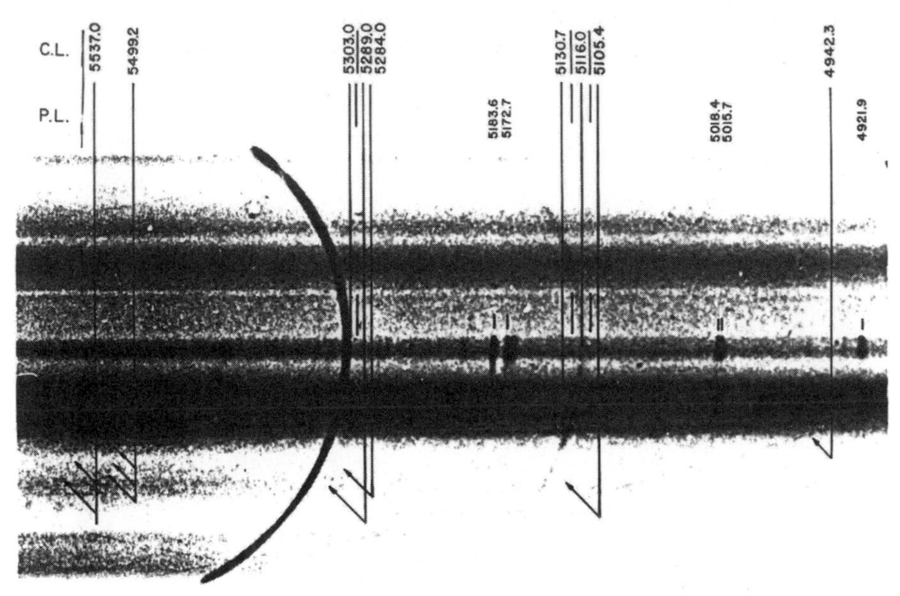

Fig. 4.21. One of the spectra of the inner corona on the Sun's western limb, in visible light, obtained by Bernard Lyot and Khairy Aly during the total eclipse of 25 February 1952 at Khartoum. (IAP/CNRS.)

and a fine streamer, very long, to the east; magnificent polar plumes.' Lyot's experiments were a total success, and the programme was carried out in its entirety. Three minutes and nine seconds later, the Sun's light burst forth again. At Fort Stanley, the radio astronomers continued their recordings, as the radio Sun is wider than its visible counterpart, with coronal emissions being registered.

It is of interest to note that coronal radio emissions had already been recorded, for the first time, a few years earlier, at the end of World War II. For the eclipse of 20 May 1947, Soviet scientists, on a costly and prestigious expedition aboard the vessel *Gribojedov*, off Brazil, intended to pursue such research. The 10 × 8-m radar antenna used on the ship was of American origin. Young, brilliant scientists such as I. Shklovskii and V. Ginzburg took part in this expedition, and were the first to investigate the 'non-thermal' nature of the solar corona. Ginzburg was to become one of the great specialists in cosmic rays, and Shklovskii went on to become Russia's best-known astrophysicist.

Lyot developed his images, and prepared a contact print for the Paris Observatory. However, the rather one-sided agreement with the Egyptians meant that he was not allowed to keep the fruits of his labours. Professor Madwar and his assistant Khairy Aly suspected that Lyot had made major discoveries, and hoped that some recognition by the international community might also devolve upon the Egyptians.

Professor Madwar would allow Lyot to make use of spectrograms at Helwan Observatory, near Cairo, and produce a joint paper with Khairy Aly for the next meeting of the International Astronomical Union, which was scheduled for September 1952 in Rome. Lyot took on this work, measuring wavelengths of numerous new spectral lines; but, sadly, he died of a heart attack on 2 April 1952, worn out by two years of continuous effort and by the Egyptian climate. Madame Madeleine Lyot, who had supported her husband throughout the expedition, finally managed to obtain the repatriation of his body to France, after many attempts to cut through bureaucracy. Académicien Lyot was buried at the Père-Lachaise cemetery on 22 April.

André Danjon, at the time director of the Paris Observatory, requested the return of the original photographs from the Egyptians, since Lyot's contact images did not lend themselves to close microphotometric scrutiny. The Egyptians refused, claiming that the images were Egypt's property, but, after negotiations and mediation by the International Astronomical Union's M. Minnaert (then director of Utrecht Observatory), they agreed to lend them to the Meudon Observatory for one year. A rapid, preliminary qualitative analysis, entrusted to Audouin Dollfus, revealed four spectral lines never before observed. The results were published at the Académie des Sciences in November 1952. Follow-up work was undertaken at the Institut d'Astrophysique de Paris by Charlotte Pecker and Lucienne Divan, with much material for analysis at their disposal. The exceptional work carried out by Lyot and his colleagues was finally published in its entirety in 1958.

Lyot's concentric-slit spectrograph formed the basis of many later eclipse experiments. His inventions have earned for him the title of France's most famous twentieth-century astrophysicist, and he did much to further the progress of astronomy with his work on the atmosphere of the Sun.

Eclipses, 1960–1980

These have been the most fruitful years for scientific findings, with an abundance of observations carried out on numerous eclipse expeditions, even if only photography was the aim. Only in 1991 were CCD cameras first used for eclipses, although mention should be made of observations by a vigorous group from the Paris Observatory, led by P. Felenbock and R. Michard, who used a prototype electronic camera, inspired by the work of A. Lallemand, to obtain monochromatic images of coronae between 1970 and 1973.

Interest in the study of the corona intensified during these years. 'Space-age' studies were still in their infancy, boosted by the great success of Skylab (1973), and other space missions. The colossal investment in space research and in the investigation of the interplanetary medium began to bear fruit.

The question of the dust component (F corona) remained unresolved, however, and a few somewhat speculative theories were put forward. Predictions included that of thermal rings, or thermal shells, around the Sun, due to infrared emission by accumulations of dust in its vicinity. The composition of these entities was hotly debated, and the altitude most often suggested for them was four solar radii. Although the whole idea has been more or less abandoned nowadays, several eclipse studies published in the 1980s refer to it. The falsity of this idea remains to be proved. This involvement of interplanetary dust with purely coronal radiation still

Fig. 4.22. Image processed to bring out coronal plasma structures. At centre, the soft X-ray corona, in negative, from Skylab observations on the same day (AS & E). The original white-light corona image was taken in Chad on 20 June 1973, on Ekta 100 with a 150-mm diameter neutral-density radial-gradient filter, exposure 20 s. A: a large streamer face-on; B: the well-defined edge of a large streamer; C: coronal hole at the north pole. (J. Fagot and S. Koutchmy, IAP/CNRS.)

causes confusion in the interpretation of polarimetric data; furthermore, purely spectroscopic data still seem to indicate the presence of a 'cold' gaseous corona. What we have here, in fact, are undesirable parasitic effects, due to doubly scattered light, showing how difficult it can be to correctly interpret 'deep' eclipse observations.

So deep is thc infatuation with eclipses among scientists, that some teams are even preparing experiments to re-verify Einstein's predictions concerning the bending of light rays – perhaps a rather less significant activity these days, when other more precise verifications have been made. Scientists are measuring the dimensions of the Moon or even the shape of the Sun, with judicious use of ephemeris data, and some are investigating the geophysical effects (for example, the propagation of gravity waves induced in the Earth's atmosphere) caused by the passage, at supersonic speed, of the Moon's shadow across the Earth's surface.

It seems, with hindsight, that progress made in these decades was due to precision photometry, including 'absolute' photometry, and to polarimetric measurements; and of course, to spectroscopy of the corona and the transition region, including the chromosphere.

One aspect illustrating the progress made during this period was the new science of spectropolarimetry. Early measurements, dating from 1954, were published in 1960 by E.I. Mogilevsky and G.M. Nikolsky, of Izmiran (Moscow). Their ground-

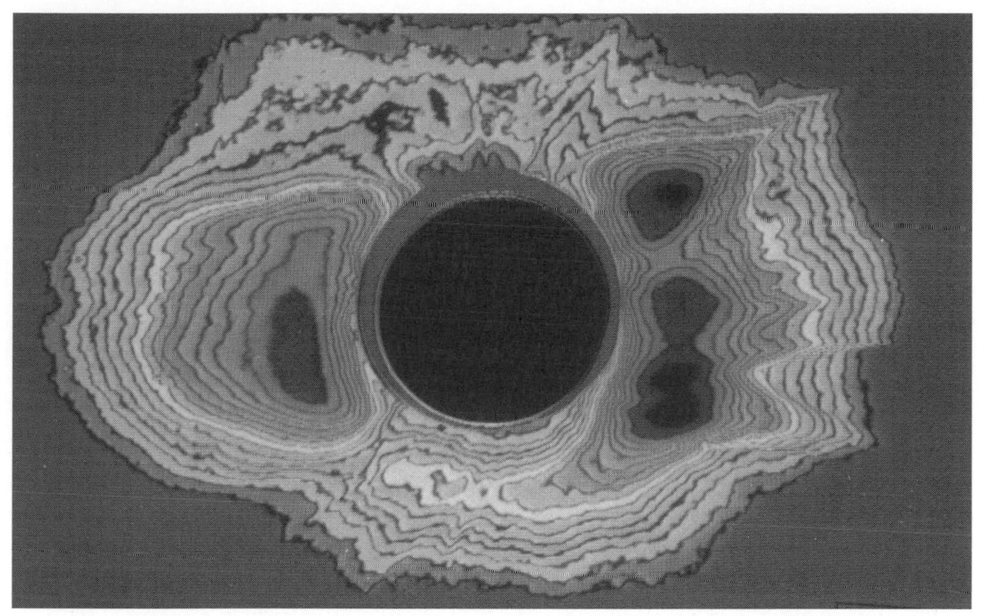

Fig. 4.23. Pattern of linear polarisation contours in the corona, at the total eclipse of 30 June 1973. The darkest areas, to the west, indicate amounts slightly above 45%; at the periphery, the amount of polarisation is about 7%, with intermediate values in between. (IAP/CNRS.)

breaking paper caused a stir. They announced strong polarisation – from 20% to 54% – of the forbidden green line of FeXIV in the inner corona, and proposed explanations. To be fair, it is worth remembering that, little more than ten years afterwards, an American team (J.M. Beckers and W. J. Wagner, *Solar Physics*, 1971), published some startling findings, based on results obtained by a different method during the 1970 solar eclipse: for the same line, polarisation was essentially nil (< 1%)! The true value evidently lies somewhere between these two extremes, but much more experimentation will be required to clarify the situation, which will doubtless still figure on the agenda in the twenty-first century. Fortunately, theory has advanced over the years, and the phenomenon of polarisation of the lines, and its links with the mechanism of the formation and effects of the magnetic field, is now quite well understood. Noteworthy in this field has been work by P. Charvin (formerly president of the Paris Observatory); the thesis by J. Arnaud (Pic-du-Midi Observatory, Toulouse); and the theoretical work of S. Sahal-Bréchot (Meudon Observatory).

This problem, the resolution of which would somewhat increase our understanding of the contribution of the coronal magnetic field, has been confronted on various occasions, through polarimetric and monochromatic imaging: a very narrow interference filter is used which selects a forbidden line (usually the green line of FeXIV, with its great intensity) to form an image of the inner corona through a linear polariser, working in at least three directions of polarisation. It has been shown in this way that at least four positions are needed to achieve reasonable accuracy. J. Sykora (Slovakia), using this method, has secured excellent images of the corona in both the green line of FeXIV, and the red line of FeX (637.4 nm), the latter not being polarised. Nevertheless, interference filters are often 'corrupted' by strongly polarised and continuous radiation from the white-light corona, and its subtraction is always a tricky business. This, however, should be rectified at some time in the future, due to advances in the modelling of the K corona, based on all the photometric and polarimetric work carried out between 1960 and 1980.

One of the most significant eclipses of the beginning of this period is that of 15 February 1961, which was well recorded across southern Europe, from the south of France, through northern Italy and Yugoslavia, and onward into the former Soviet Union, though the weather in the Crimea was cloudy. Meudon Observatory sent teams to Hvar (former Yugoslavia) and to the Crimea. In spite of attempts by aircraft to break up the clouds by chemical seeding before totality, the Crimean team came home empty-handed. For their part, the Hvar team enjoyed excellent conditions, and unique spectra were obtained, with equipment of some considerable dimensions, including two spectrographs for ultraviolet and visible light. P. Sotirovsky managed to obtain a very detailed prominence spectrum, which was published in *Annales d'Astrophysique* in 1965, and still serves as a reference, with its impressive array of very faint lines. Sotirovsky and R. Michard also undertook analyses of polarised images of the white-light corona, concentrating for the first time upon careful measurements in the outer corona, dominated by the dusty F corona. They also measured effects of stray light in the line of sight, originating in reflected sunlight beyond the Moon's shadow, and illuminating the horizon during totality.

A surprising aspect of this effect, arising from doubly-scattered stray light, is that it is even more invasive in the case of airborne eclipse observations: at high altitudes, the brightness of the horizon more easily intrudes into the line of sight. This effect probably lies behind the discovery, by various observers during eclipses, of extra lines of chromospheric origin, which have fuelled speculation about a cool component of the corona. Another unwanted effect is that caused by irregularities at the lunar limb, often in combination with the Moon's passage in front of fine coronal structures, interfering considerably with interpretation of the so-called 'flash' spectra. Accurate interpretation here depends upon the Moon's edge successively obscuring different layers of the chromosphere and the transition region with the corona, while spectra are taken at a rapid rate (at least one or two every second). This procedure has to be carried out with a fairly wide curved slit, or even without a slit, which allows investigation of a wide segment of the Sun's edge.

What was indisputably the finest set of flash spectra ever obtained came from the eclipse of 5 February 1962, in which several American observatories and laboratories (Boulder, Washington and Hawaii) were involved. The instrumentation, designed especially for this eclipse, was produced from 1958 onwards by R.B. Dunn at Sacramento Peak, and the results were published in a lengthy and now classic paper in 1968 (*Astrophysical Journal, Supplement*). The whole range of wavelengths from 320 to 910 nm was covered, and each spectrum measured 60 × 15 cm! More than

100 spectra of this size were obtained. This spectrum of the 'extended' chromosphere has still not yet been completely interpreted, and models of the transition region since proposed in the United States still do not refer to it. There is, however, one less well-known study, by T. Hirayama (Tokyo Observatory), which was inspired by these results, and by others from Japanese sources, most notable among them being the work of E. Hiei (former director of Tokyo Observatory's Solar Division). The question of chromospheric extensions into the corona is present everywhere in eclipse observations, and observers from Russia, France, Japan, America and other countries, including, more recently, India, have all applied themselves to discovering what goes on in the zone known as the transition region. To this end, G.M. Nikolsky, I. Kim and the IAP/CNRS team prepared several experiments for the 1981 Kazakhstan eclipse. Joint use was made both of spectroscopic methods – including the Fabry–Pérot interferometric procedure – and of photography. But the key to the problem lay in the possibility of obtaining excellent spatial resolution, and in the comprehension of small-scale dynamic phenomena, with the solution demanding good temporal and spectral resolution for measuring Doppler effects. In this way, and from observations like those made in 1981, the discovery of small plasmoids, in which hydrogen is almost wholly ionised, was rendered progressively more likely during the years in question.

Some extremely interesting observations of coronal spectra were attempted during the eclipse of 15 February 1961, when, remarkably, totality was observable from France's biggest astronomical observatory, at Saint-Michel-de-Provence (CNRS). The great 193-cm reflector was used by the director of observations, C. Fehrenbach, and G. Wlérick from the Paris Observatory played an active part. Their goal was to obtain spectra from selected sites of the inner corona, with a radial slit more than 1 arcmin long. For practical reasons, the Sun's south pole was chosen. The spectral domain was also new, in that it included the near-infrared. Several coronal lines were recorded, among them an FeXV line near 706 nm, showing that, even in polar regions, temperatures may be very high within certain structures (in this example, more than 2 million K). Spatial resolution along the slit was obviously very good. The best exposure, of almost 50 seconds, reveals numerous structures intercepted by the slit, but none has been identified.

Other observations carried out at this observatory included those of D. Chalonge (IAP/CNRS), whose team used the 80-cm Cassegrain refractor with Chalonge's renowned small UV spectrograph, with additional visible-light capability. A series of spectra resulted, and the aim was to study the colour of the inner corona and its polarisation in the spectrum. Results obtained with the 120-cm photographic telescope proved even more spectacular: J. Bigay succeeded in capturing an image of the inner corona on a photographic plate – possibly the best image of the inner corona ever taken, with a resolution of the order of 2 arcsec almost everywhere. All these results were published in an extensive article, written by American astrophysicist D.H. Menzel (director of the Astrophysics Center at Harvard University) for *Sky and Telescope* (**21**, 1961). Menzel himself had set up an observing site in Italy, and tried out the recently introduced film process invented by Land, using the new Polaroid camera. No photograph of the corona had ever been

Fig. 4.24. Preparations for the solar eclipse of 31 July 1981, in Kazakhstan. At left is G.M. Nikolsky, one of Russia's best solar physicists; behind is G. Stellmacher. (IAP/CNRS.)

Fig. 4.25. Photograph of the corona taken by M. Laffineur, M. Bloch and M. Bretz at the Haute-Provence Observatory during the eclipse of 15 February 1961. The equipment included a 15-cm camera of 3.6 m focal length, with a focal filter at 610–645 nm, and a rotating diaphragm producing radial attenuation of brightness from the limb out to 1.5 solar radii. This photograph has played an important part in the evolution of our ideas about the plasma corona, as it shows structures very clearly, due to the compensation technique for the radial gradient. (IAP/CNRS.)

Fig. 4.26. The quadruple eclipse telescope developed by the Institut d'Astrophysique de Paris/CNRS for the eclipses of 1968 and 1970, equipped with radial-gradient neutral-density filters and radial and tangential rotating polarisers. The instrument is shown at the observing site near Oaxaca, Mexico, on 7 March 1970. (IAP/CNRS.)

obtained with this type of camera, neither by Menzel nor by G.E. Moreton (then director of the well-known Lockheed Solar Observatory), who observed with him. Suffice it to say that time will have to pass before the 'instant development' camera – though ideal for many other uses – secures good images of eclipses.

The *Sky and Telescope* articles mentioned feature some striking images which challenged some of the received wisdom in solar physics. They were obtained by Laffineur and colleagues from Haute-Provence Observatory, and show the more extended corona, with structures of the plasma corona, after albeit imperfect compensation of the mean brightness gradient. One particular type of radial-gradient neutral-density filter was used near the focus of the horizontal refractor–coelostat, set up on the observatory terrace.

The same method had been employed in 1937 by an American team, who placed near the focus a rotating device with two small heart-shaped wings, which avoided overexposure of the internal parts of the corona.

With such advanced photography, M. Laffineur was successfully able to contribute to the modification of theories on the corona: now, it was obligatory to take into account the influence of the magnetic field when modelling the corona, and the physics of coronal structures became a much-discussed theme. Not long afterwards, during the eclipse of 1962, K. Saito of Japan took up the idea again; this time the rotating device had wings reminiscent of a dragon. The best example of the use of this method was that of G. Newkirk (director of the High Altitude

Observatory at Boulder), who constructed the 'coronal camera' which still bears his name. This camera, still with the same instrumental parameters, secured a great number of eclipse images with the addition of a veritable radial-gradient neutral-density filter. The first of these was taken at the time of the Bolivian eclipse in 1966; later images date from 1970, 1973, and afterwards.

Almost simultaneously, Laffineur had begun work on his new 'quadruple' camera, with which one of the authors obtained some unique images in 1968 and later in 1970, chiefly in axially symmetric polarised light (all other analyses of polarisation use linear polarisers). It was indeed difficult to interpret the results, in the light of hypotheses aimed at understanding 'abnormal' polarisations and the importance of very energetic electrons in the corona. To date, these measurements remain unconfirmed, and the interpretation of delicate polarisation measurements remains decidedly open.

With the same instrument, during the 1970 eclipse, the IAP team obtained the first 12.5 × 10-cm colour image through a radial-gradient neutral-density filter. It was, however, never to be acclaimed, as the original negative went into the incinerator of a large New York hotel. While we were on our way back from Rochester, where we had been to assure careful development of the film at Kodak, a chambermaid found the negative in its 'Mexican' box (which was in an undeniably frayed condition) on

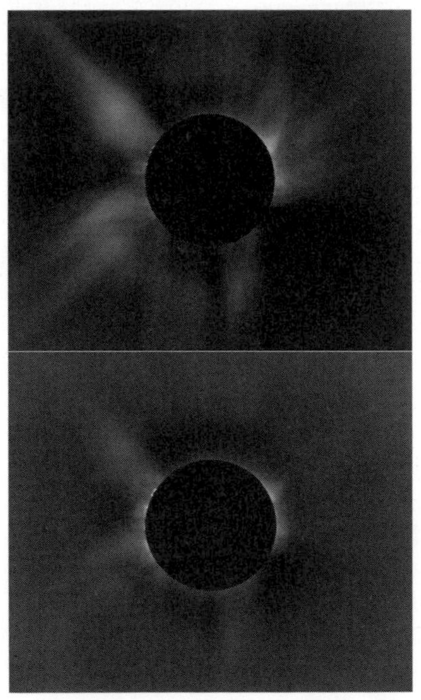

Fig. 4.27. Images of the white-light corona, observed in white light polarised tangentially to the solar limb (above), and radially (below). Photographs from the Mexican eclipse of 7 March 1970, taken by the IAP/CNRS team with the quadruple telescope. 100-mm Clavé objective, focal length 1,550 mm, 12.5 × 10-cm black-and-white film, Ekta Pan 100 ISO, exposure 60 s. (IAP/CNRS.)

the bed, and threw it away while cleaning the room, at the same time as we were in Manhattan celebrating our success! Fortunately, we had already sorted the many black-and-white images due for our attention, and they were already packed away.

In 1971 the Institut d'Astrophysique de Paris began development of a new coronal camera, of focal length 3 m, with a 20-cm 'standard' Zeiss objective salvaged from the *Carte du Ciel* project of the 1920s. Another initiative was the construction of an eclipse coelostat with a 30-cm Cervit mirror. With this camera, what are probably the best images of the extended corona have been obtained, due to its 150-mm radial-gradient neutral-density filter (from MTO, France) and its use of 24 × 18-cm colour film.

In 1973, collaboration with the University of Kiev led to a notable 'first': by comparing two images of the same eclipse taken at two different locations (Mauritania and Chad), it was possible to discover the origin of the so-called 'slow' solar wind. Small plasma clouds are continually detached, by magnetic instabilities, from larger streamers, and propagate outwards through the corona at velocities greater than 100 km s^{-1} into interplanetary space. This discovery was announced in

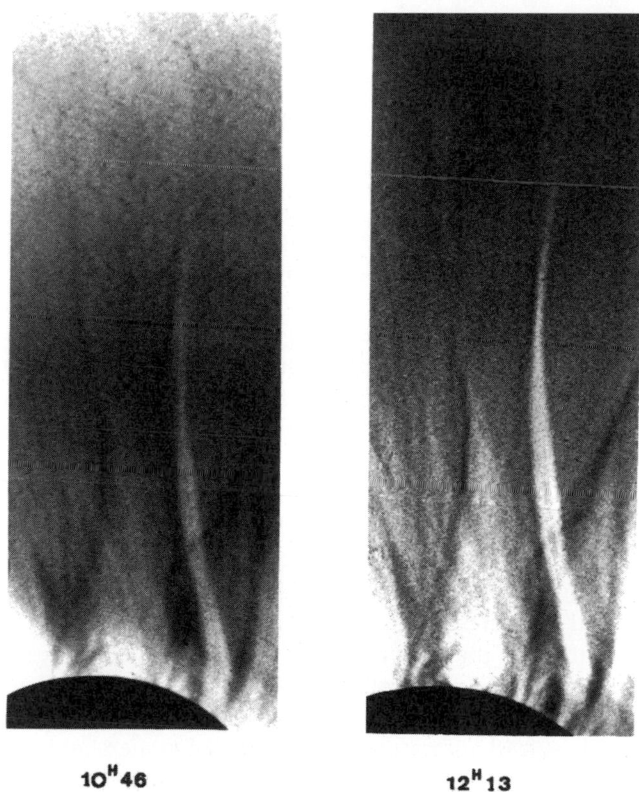

Fig. 4.28. This comparison of two images taken during the eclipse of 30 June 1973 shows how a plasma cloud detaches itself from the edge of a large streamer. (IAP/CNRS and University of Kiev, Ukraine.)

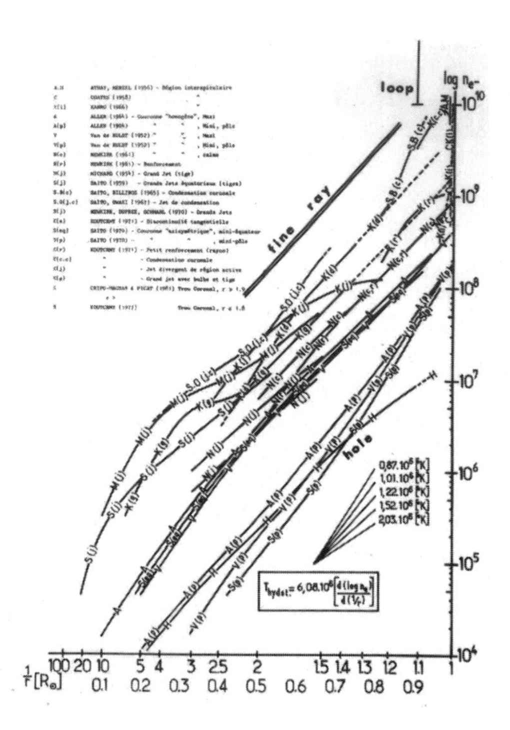

Fig. 4.29. Electron densities in the plasma corona, derived from photometric and polarimetric observations by various authorities between 1960 and 1980. Densities – as number of electrons per cubic metre – are plotted against radial distance. The slope of these curves is determined by the so-called hydrostatic temperature, and also by the radial velocity field. (IAP/CNRS.)

Nature in 1973, and was later (more than a quarter of a century later, in 1997!) definitively confirmed by much more detailed observations from one of SOHO's coronagraphs. Not all observations are similarly supported, partly because they are not sufficiently analysed, and are forgotten. Thousands of pages have been published on eclipses from 1960 to 1980, and it is impossible to mention them all, and pointless to try. Indeed, a case in point is the work of Waldmeier, of the Zürich Observatory, a veteran eclipse observer who never missed one during his long career. Little is seen nowadays of his many reports, which always followed the same format, with thoroughgoing coronal drawings and isophotic contours.

From the vast numbers of photometric and polarimetric eclipse results, we have learned, for example, about the mean radial distribution of electrons in the plasma corona. The interpretation of these data is now well accepted, and the results are firmly established. They lead us naturally towards an understanding of 'global' coronal physics, although they are less instructive about the physics of structures. Greater resolution is therefore needed, as well as more sophisticated investigation. As technology advances, we shall no doubt carry out more and more accurate eclipse observations, and new discoveries await us. We can see, in the light of past experience, that the road to them will not be an easy one.

Fig. 4.30. The corona during the eclipse of 30 June 1973. This montage, realised through different photographic treatments, reveals the Sun's disk in Hα, and details of the innermost and outermost coronal regions. (IAP/CNRS and C. Keller, Los Alamos Laboratory.)

Total eclipse over Mauna Kea

On 11 July 1991, on the island of Hawaii, solar physicists had the amazing good fortune to be able to turn some of the best telescopes on the planet towards a total eclipse of the Sun. The last total solar eclipse of the twentieth century to be seen from American territory would sweep across the observatory, situated 4,200 m high on Mauna Kea. Numerous teams from all over the world geared themselves up to record this event, for which there was a good choice of excellent observing sites. Many chose to go to Mexico or Brazil. Here was an opportunity for teams from different nations to pool their images of the corona, taken with similar equipment (using a 15-cm radial filter at 3 m focal length) from widely spaced sites, covering several hours of observation. The process went well, effectively coordinated by J. Zirker, who created the acronym MICE (Multi-site International Coronal Experiment). Several well-equipped Japanese teams set up their instruments in Baja California (Mexico). For the first time, many CCD cameras were in use, with very narrow interference filters. Mauna Loa and Mauna Kea, in Hawaii, attracted numerous observers with experiments for the large instruments there, and with projects specially designed for the occasion. The major Boston-based public television channel WGBH invested more than a million dollars in the making of a popular film about the event. Three years' work went into its production, and it was shown several times in the United States, where this 'eclipse of the century' excited much interest. Director Tom Levison had dispatched several teams of technicians to Hawaii for the eclipse, to film both the event and the scientists at work.

On the morning of 11 July 1991, atmospheric conditions were most unfavourable,

Fig. 4.31. This wide-angle photograph, taken during totality on 11 July 1991, at 4,200 m on Mauna Kea (Hawaii), shows the sky partly polluted by high-altitude volcanic dust from Mount Pinatubo in the Philippines. (IOA/University of Hawaii.)

and heavy clouds covered most of Hawaii, with its thousands of hopeful observers. Only the astronomers perched on the mountain tops, above the sea of clouds, were spared. The sky was also tainted by high-altitude volcanic dust, drifting eastwards over Hawaii from Mount Pinatubo in the Philippines.

However, several experiments carried out in Hawaii during the 250 seconds of the total eclipse led to considerable progress in our understanding of the physics of the solar corona. On that day, at 7 h 28 m, the eclipse was recorded by a 3.6-m reflector, the biggest ever pointed at the Sun, a 3-m infrared telescope, a 2.2-m telescope, two 60-cm telescopes, and lastly two radio telescopes, one 10 m and the other 15 m in diameter. A highlight of the observing campaign was the work of the CFH (Canada–France–Hawaii) Telescope, the most prestigious and powerful instrument in use. The international team on the CFH was led by a party from the Institut d'Astrophysique de Paris, and the telescope was equipped with the finest instrumentation at its prime focus. Two imaging devices, motorised to a capability of four 70-mm format images per second and using Kodak TP2415 film, and two CCD cameras recording 30 images per second, were disposed around the circle of the lunar disk, and operated remotely from the telescope's control room. The power of the CFH telescope enabled best ever spatial resolution achieved during observations of the corona. Details only 500 km across were recorded, and the speed of the cameras ensured unrivalled image quality. The fastest-moving coronal structures were measured at about 100 km s^{-1}, with possible error in measuring speeds of displacement of only 10 km s^{-1}, due to system refinements. The photographs – digitally processed with algorithms specially developed for this kind of analysis – revealed a host of structures and coronal loops, which surpassed all expectations at the time. New perspectives on the delicate phenomena of the plasma corona were opened up by these images, which clearly show the richness and complexity of the structures, suggesting essentially small-scale dynamic (flow) phenomena, and abrupt instability events on the larger scale. The whole structure of the plasma corona seems interwoven with fine structures, twisting and braiding, interacting and regrouping to form larger phenomena such as streamers. Fine structures (of dimensions less than 1 arcsec – less than 700 km) appear to be very variable, and are detected above active equatorial regions at altitudes of more than 100,000 km above the solar surface, with lifetimes of the order of 100 seconds. Larger features have longer lifetimes.

An isolated cloud of coronal plasma, or plasmoid, 2,500 km across, was filmed early on in a CCD video sequence. The plasmoid split up several times, producing short-lived filamentous structures. The cloud disappeared after 200 seconds. This observation of a coronal plasmoid and its associated dynamic phenomena shows

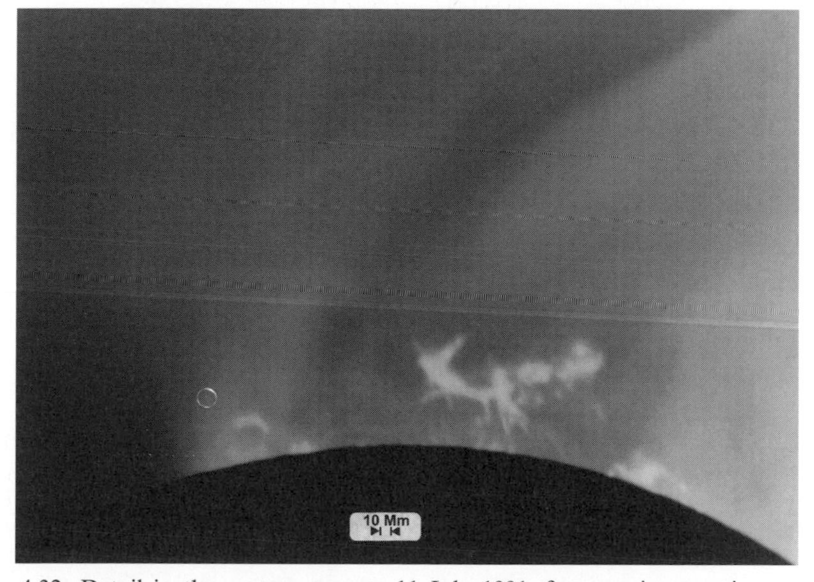

Fig. 4.32. Detail in the eastern corona, 11 July 1991, from an image using a radial-gradient neutral-density filter, and a red filter to bring out 'cool' emissions in prominences. The plasmoid observed with the CFH telescope is clearly identifiable in this photograph, taken with an instrument of 3 m focal length, objective 20 cm, exposure 20 s. (J.-P. Zimmermann, IAP/CNRS.)

coronal dynamics in action at a resolution of less than 1 arcsec. At the Institut d'Astrophysique de Paris, a study of this new phenomenon was undertaken by C. Delannée in her PhD.

It has been shown that these plasma clouds, when ejected, are at first diamagnetic; as they fade against the coronal background, it is possible that small-scale magnetohydrodynamic (MHD) phenomena are responsible for coronal heating. The wealth of data obtained by other telescopes on Mauna Kea takes us into many other domains, among them observation of interplanetary dust, detection of thermal microstructures in the corona, acquisition of high-resolution images of the solar limb at a wavelength of 12 μm and of the Sun's environment at 0.85 μm and 1.2 μm, and investigation of the extension of the chromosphere in the millimetric domain. Never before has the Sun been observed simultaneously by so many such powerful telescopes, working in so many different domains of energy. The main results of the experiments carried out with the CFH telescope are further discussed in Appendix B.

ECLIPSES IN FLIGHT

(Parts of this section have been taken from the article *L'éclipse en plein vol*, written by one of the authors for the magazine *Ciel et Espace*, July/August 1990.)

On 14 August 1937, *Time* magazine carried a piece of scientific news heralding a new era of airborne astronomical observation, in the high troposphere and stratosphere: US Army major A.W. Stevens had successfully observed the total solar eclipse of 8 June 1937 from a DC-2 aircraft flying at about 9,000 m over the Peruvian Andes. As well as drawing solar-maximum coronal streamers (see *The Sky*, 1937, article by Fisher *et al.*), Stevens announced the existence of an extensive 'aureole', out to 1.6 million km, into which the streamers poured. It is now known that this aureole is in fact the innermost part, nearest the Sun, of the zodiacal cloud. The US Navy also made a small contribution to science during this eclipse, stationing one of its ships in the Pacific in the path of the Moon's shadow, and moving with it to add a few seconds of totality. Since total eclipses often move largely across oceans, ships, including the biggest liners, are nowadays much used for observation, especially on 'eclipse cruises'. These two anecdotal 'firsts' mark the beginning of a continuing era of more than 60 years of seaborne and airborne eclipse hunting.

Why aircraft?

A total eclipse is a short-lived event. The longer partial stages hold little of interest, but when totality arrives, the observer is plunged into the deep twilight of the umbral cone, swept across the Earth by the moving Moon as a shadow from 100 to 200 km across. This cone passes rapidly across the Earth's surface at more than 2,000 km h^{-1}. The thermal shock accompanying the shadow generates gravity waves in the upper atmosphere, measured at ground level with microbarographs placed in the path of the eclipse if weather conditions (clear and anticyclonic) are right. The ionosphere, abruptly swept and disturbed by the shadow, is bereft of ionising radiation.

Astronomers do not concern themselves greatly with such phenomena; they are more interested in the circumsolar atmosphere, from its innermost regions outwards. As the Moon – with its imperfect profile due to the rugged nature of its surface – begins to cover the Sun's apparent disk, the first unseen layer of the Sun's atmosphere is revealed: the chromosphere, closest in to the solar surface, and once known as the 'reversing layer'. It has taken nearly a century to unveil the essentially dynamic and heterogeneous nature of this complex region. Its exact origin, connected with, among other things, small-scale magnetism and hydrodynamic flow shocks, remains a mystery. After totality begins, and as the lunar limb trespasses further onto the fine fringe of the chromosphere, and then onto the transition region forming the base of the corona, we can explore mysterious extensions, like knives cutting the successive layers of the Sun's atmosphere into thin slices. Also observed is a very sudden reversal in all the well-known spectral absorption lines seen when the Sun is not eclipsed – the 'fingerprints' of the many chemical elements which make up our star. Abruptly emerging from the dense layers of the solar atmosphere, these elements now produce bright emission lines. Here is unambiguous evidence of gases in an altered state, which we need to know more about, as it is very far from the thermodynamic equilibrium so easily described in our equations.

So, astronomers try to prolong those precious seconds of totality by staying within the Moon's umbral shadow, and only an aircraft flying at more than mach 2 can keep them there, extending the duration of the eclipse. The corona and its possible variations can then be unhurriedly analysed.

It is feasible to do the same thing on the ground, by spacing observing stations along the trajectory, but the decisive advantage of airborne observation is that the observer remains above the clouds, with the added advantages of sky clarity and spectral transmission impossible to achieve from down below. The higher one is able

Fig. 4.33. A photograph taken after the eclipse flight of 22/23 November 1984. The aircraft is a Guardian patrol aircraft of the French Naval Air Arm, based at Nouméa, New Caledonia. During the flight, very precise occultation timings were made, giving information about the length of chords across the lunar disk.

to go in the atmosphere, the darker or 'deeper' the sky looks. Transmission of infrared and extreme ultraviolet radiation is increased, and new observing windows are opened.

Early post-war eclipse flights

Towards the end of World War II, the eclipse of 9 July 1945 was seen from central Canada. Four Canadian aircraft were used for a pioneering set of observations – an experiment repeated over Labrador during the eclipse of 30 June 1954.

In 1951 the US Air Force made available a B-29 bomber, with its pressurised cabin, to a group of physicists from the University of Colorado. This aircraft made the first near-stratospheric eclipse flight at an altitude of 9,600 m.

On 25 February 1952, the Moon and the Sun met again, over Mecca. This eclipse – probably the first of the modern era of eclipse observation, with its stratospheric flights, radioelectronic measurements, complete coronal spectrum, and so on – occurred shortly before the tragic death in Egypt of the great French astrophysicist Bernard Lyot, who had carried out splendid observations of the event.

The B-29 flight was the first of many such ventures, and the corona could now be studied to its maximum extent, several degrees from the Sun. It had long been suspected that there must be some continuity between the far extensions of the corona and the zodiacal light, which extends for tens of degrees along the zodiac, as observed from the best dark sites at the end of twilight, when the Sun is far enough below the horizon for the stars to appear; the zodiacal light decreases in intensity with distance from the horizon. It is noticeably absent very close to the horizon, its light being absorbed by the lower layers of the Earth's atmosphere. It is now known that this light is due to sunlight being diffracted and scattered by the tiny particles of the zodiacal cloud, orbiting the Sun and spiralling in imperceptibly as the pressure of solar radiation decelerates them (aberration of light). This same solar radiation, in its 'harder' forms of UV and X-rays, is also capable of ionising interplanetary dust grains, electrically charging them and thereby submitting them to electromagnetic forces. Astronomers have long wondered what relationship exists between the corona, traversed by large ominous streamers, and the mild glow of the interplanetary medium. To resolve the problem, it has been necessary to take very careful measurements. By this, we mean the measurement of luminous intensities 1,000–100,000 times fainter than those of the inner corona during totality. The screening effect of the Earth's atmosphere is a hindrance even when the air is clean. Conditions required are similar to those for observing the corona of the uneclipsed Sun with a Lyot coronagraph. For an event as ephemeral as an eclipse, an aircraft perhaps offers more than does observing from a high mountain. However, there remains the problem of the degradation of observations made through an aircraft window, which can never be as clear as the lens of a coronagraph. So, the Plexiglas dome through which the 1952 images were taken could not ensure precise measurements. The problem was partly resolved by simply removing the window, avoiding condensation and even icing problems; during the total eclipse of 30 June 1954, in a Lincoln bomber capable of an altitude of 13,000 m but flying at 9,000 m, D.E. Blackwell, from England, took observations through an open door! The following eclipse, on 20 June 1955, saw R.

Michard of the Paris Observatory successfully observing over Vietnam aboard a brand new Air Force Nord 2501 aircraft.

The unique advantage offered by airborne operations – that of the possibility of seeing the corona projected against a sky less bright than is seen from the ground – had been definitively exploited. In the latter case, the sky was not completely clear – there were cirrus clouds present and the non-pressurised aircraft was flying at its ceiling of 8,100 m. Only stratospheric altitudes can guarantee the absence of clouds; but then, windows are required. It should be noted that, at the time, the nature of the dynamical/solar wind aspects of the corona had not been established, and a certain confusion (still with us after 40 years!) existed as to the interpretation of the diffuse optical component: the dust component, possibly polarised, in the case of its smallest particles and of those out to a distance of a few solar radii; or mass ejections of coronal plasma, and even of great prominences reaching out into the interplanetary medium.

The aircraft comes into its own

The use of aircraft during eclipses became more common in the early 1960s. The memorable eclipse of 15 February 1961 saw some important flights; for example, by a Caravelle over Italy, and a Soviet bomber with a payload of infrared spectroscopy instruments over the USSR (V.G. Kurt, *Soviet Astronomy*). The most impressive experiments, however, mounted by various agencies and aerospace companies, took place in the USA. Undoubtedly the greatest armada of jets ever to fly beneath an eclipse took to the air over northern Canada on 20 July 1963. As well as the NASA

Fig. 4.34. A DC-8 equipped by NASA as a stratospheric laboratory for the eclipse of 1963. Note the windows placed for upward viewing. (NASA.)

expedition, whose stratospheric laboratory was installed in a DC-8 (later, in 1965, NASA similarly equipped a CV-990 aircraft), there were observers aboard two large KC-135 transports, and two F-104D jet fighters with one observer in each.

The Royal Canadian Air Force flew several aircraft with British scientists on board during the 1963 eclipse, including a CF-100 and a Yukon turbo-prop. The University of Alaska used a B-17. Experiments of many different kinds were flown, one of which was French. Some sat upon gyrostabilised platforms, and the most sophisticated experiments involved spectroscopy – notably that of R.B. Dunn, whose fine spectrograph experiment was inspired by Lyot's remarkable work in 1952. This instrument, of which two were built, flew several times and provided numerous spectra, some of excellent resolution with many lines, and of photometric quality. Françoise Magnan-Crifo carried out a thorough analysis of these. Outstanding results came from further flights during the eclipse of 30 May 1965.

Special mention must be made of the flights operated by the Los Alamos (USA) group of laboratories, which carried out perhaps the most impressive work with an epic series of stratospheric eclipse observations between 1965 and 1980. All this began in 1963, with the signing of the treaty banning nuclear tests in the atmosphere, when the USA decided to equip three large Boeing 707-type jets as surveillance

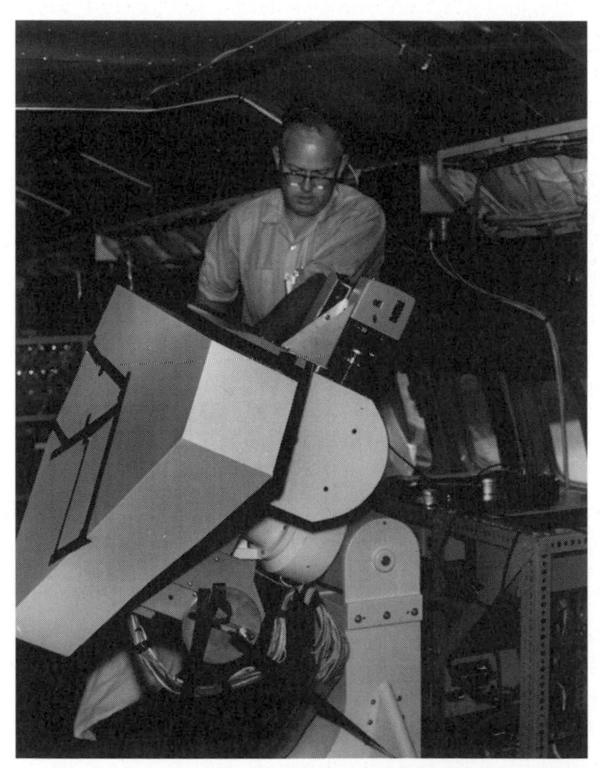

Fig. 4.35. Astrophysicist R.B. Dunn, of Sacramento Peak Observatory, with his airborne spectrograph used during the eclipses of 1963 and 1965. (NASA.)

Fig. 4.36. Part of an experiment by Los Alamos laboratories, mounted behind special lateral windows in an eclipse-chasing aircraft. Note the dry-air blower behind the window to prevent icing during the high-altitude flight, and the gyroscopic tracking system. (Los Alamos Observatory.)

aircraft to police the 'good progress' of this agreement (perhaps optimistically, given the mood of the times). Since these aircraft were kept at operational readiness, A. Cox, a Los Alamos astrophysicist, suggested early on that one of them – equipped as it was for sophisticated imaging – be used for eclipse flights. The first such expedition occurred over the Pacific, at more than 12,000 m, during the eclipse of 30 May 1965. Later, with more refined techniques, D. Liebenberg and C. Keller obtained results of great scientific value. Mounting instruments on inertially stabilised platforms, they secured unique images of the corona, showing extended coronal streamers out to more than 12 solar radii: results of which Major Stevens could scarcely have conceived in his DC-2 beneath the eclipse of 1937.

A vast mass ejection, larger than the solar disk itself, was observed from an aircraft over the Indian Ocean on 16 February 1980, with the Sun at sunspot maximum. Enormous plasma clouds were seen, in polarised light, to tear themselves away from the corona, confirming observations made by small orbiting coronagraphs with external occultation. Spectroscopic work was also undertaken, with limited success. During the eclipse of 30 June 1973, many famous scientists made these flights, heralding the systematic use of aircraft for astrophysical research. NASA's 'flying observatory', mounted in a Convair 990 in 1965, was but the first of many flights undertaken in its name. After the tragic loss of the second of these installations, Galileo II, more recent observations took place aboard the Kuiper Airborne Observatory, a C-141 transport. Capable of altitudes of more than 13,500 m, it carried a 91-cm telescope, with which mostly infrared and submillimetric work was done. During its flights in the 1980s, Jefferies, Lindsay and others measured the extension of the chromosphere into the lower layers of the corona, aided by light curves in the millimetric and submillimetric domains of radiation taken while the Moon's limb moved across. These measurements led to a model of the transition region, with its spicules. From the ground, however, spicules are essentially observed in Hα.

The epic flight of Concorde 001

Many scientific flights have taken place since 1952, but none has had the resounding success of the historic excursion of Concorde 001 on 30 June 1973. Due to its supersonic capability, it was able to provide 74 minutes of totality – a record never since equalled, and very unlikely to be – for the various teams on board, each with its specialised equipment. An essential element in these experiments was the provision of optical-quality windows allowing astronomical observations. A few years before, this was one of the factors which had spelt failure for an American initiative: for the eclipse of 7 March 1970, American astronomers, chief among them J. Pasachoff and A. Cox, had requested the use of the famous SR-71 Blackbird reconnaissance aircraft – holder of records for altitude and speed, and able to fly at more than mach 3. The services of this aircraft – which at the time was top secret – were not authorised by the Defense Department, nor by the builders, Lockheed.

We had much better luck at the time, as the prototype Concorde 001 had just emerged from its years of tests, and could reach mach 2 at stratospheric altitudes. Permission was sought to insert optical-quality windows in its fuselage, and under P. Léna, an experienced group involved in airborne infrared astrophysics at Meudon Observatory began preparations for the venture six months before the date of the eclipse. One of the authors, Serge Koutchmy, was in charge of experiments aimed at photographing the corona and its environment.

Aérospatiale (Toulouse) took a great interest in the project, and was extremely cooperative; however, considerable funding would have to be sought. Contributors to the project included the CNES (Centre National d'Etudes Spatiales) , the CNRS and the DGRST (Délégation Générale à la Recherche Scientifique et Technique). The lack of time available for such a far-reaching project was compensated for by the enthusiasm and skills of the participants. Since only the Americans had had profitable experience of eclipse flights, a group from Los Alamos was invited to take part. The Meudon team set up an ambitious infrared experiment, with a view to directing an infrared spectrograph at dust grains near the Sun, through a composite

Plate 1

Plate 2

Fig. 1 (top)
Fig. 2 (centre left)
Fig. 3 (bottom)
Fig. 4 (centre right)

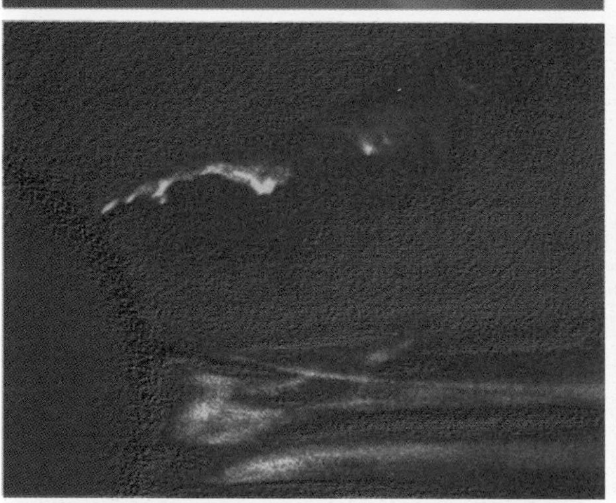

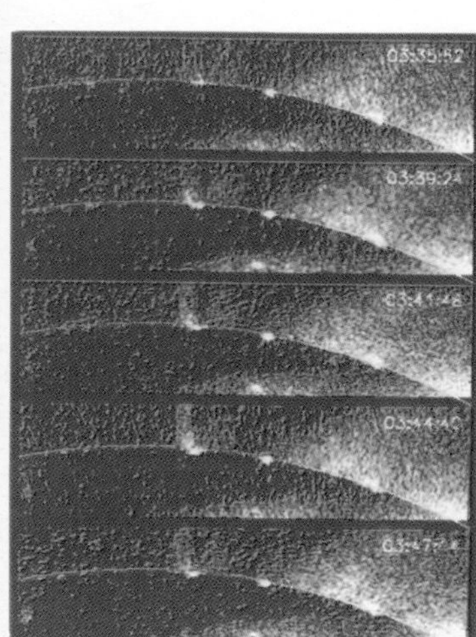

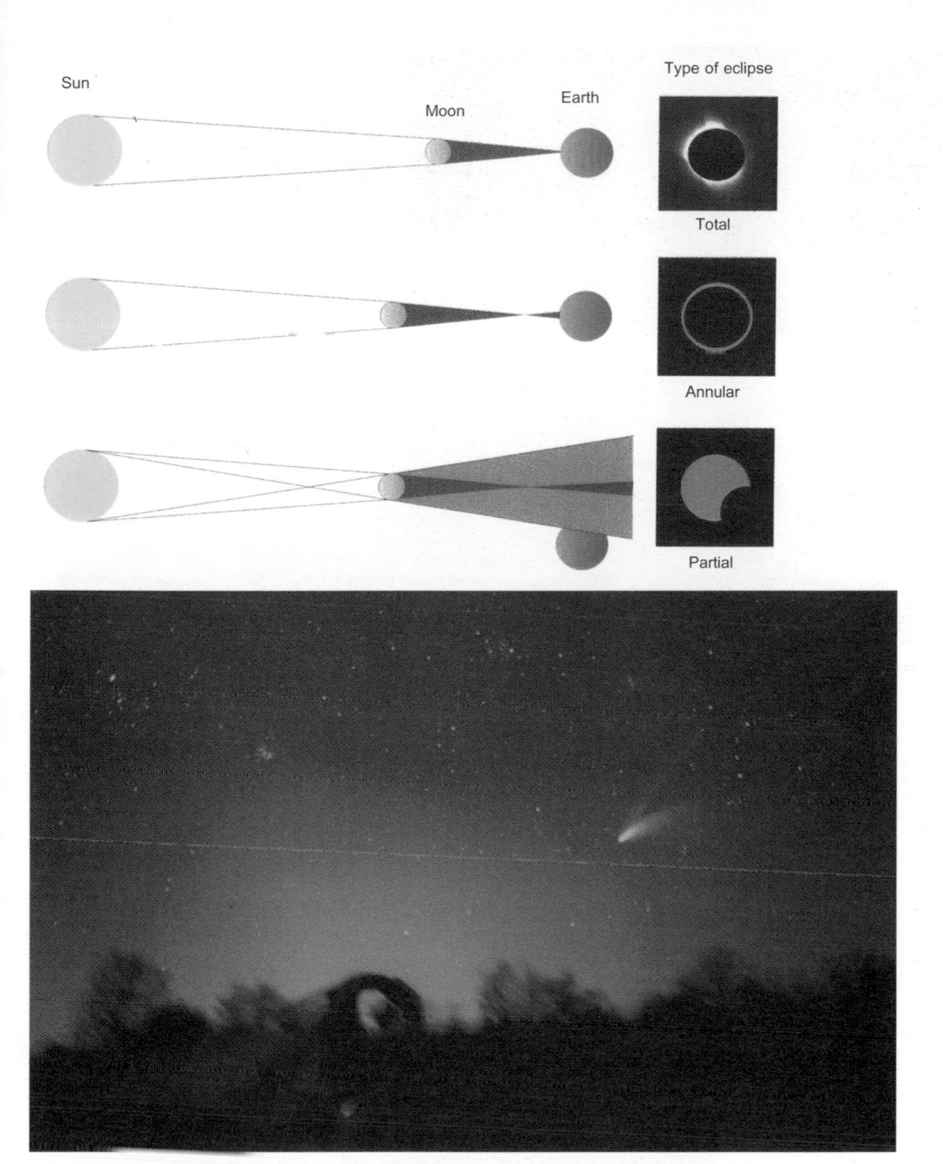

Plate 3

Fig. 1 (top)
Fig. 2 (centre)
Fig. 3 (left)

Plate 4

Fig. 1 (above)
Fig. 2 (left)

Solar Eclipses

Central

- Total
- Annular
- Annular-Total

Non central

- Partial
- Total or annular

Lunar Eclipses

- Total
- Partial
- Penumbral

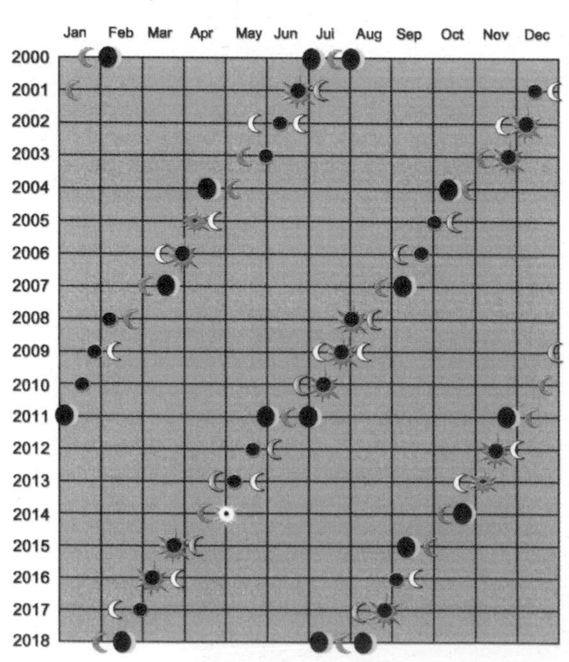

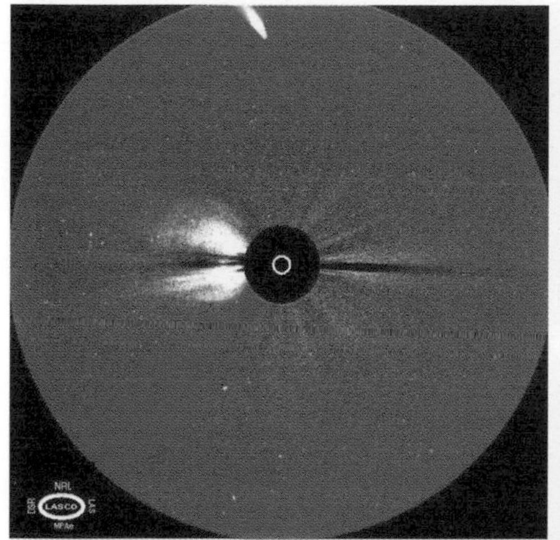

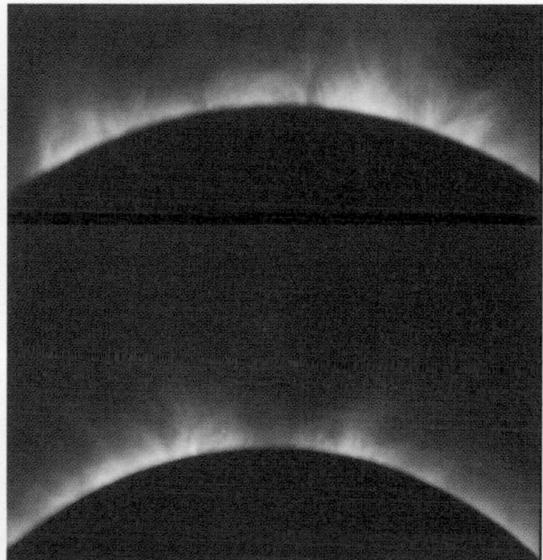

Plate 5

Fig. 1 (top left)
Fig. 2 (top right)
Fig. 3 (left)
Fig. 4 (bottom)

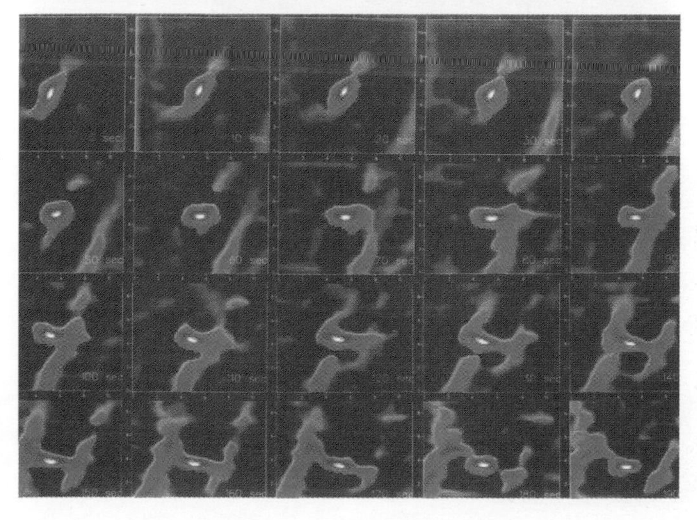

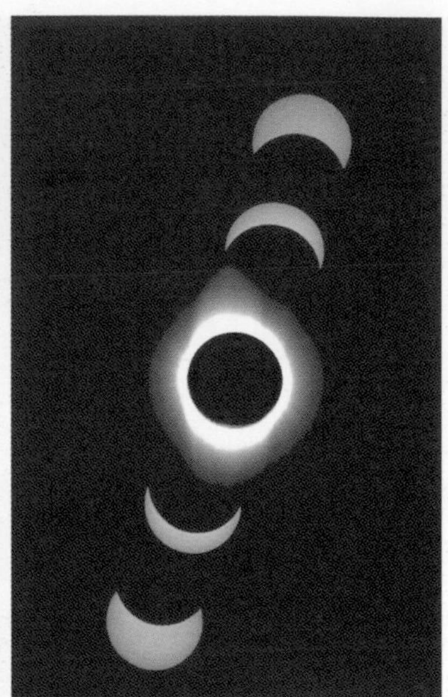

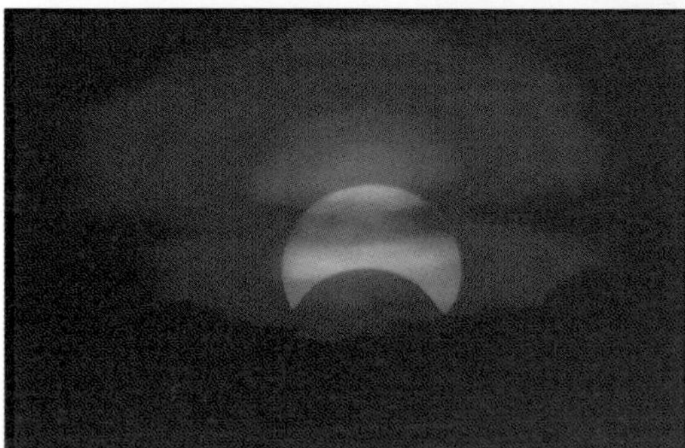

Plate 6

Fig. 1 (top left)
Fig. 2 (top right)
Fig. 3 (centre)
Fig. 4 (left)

Plate 7

Fig. 1

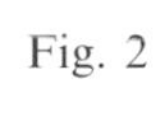

Fig. 2

Fig. 3

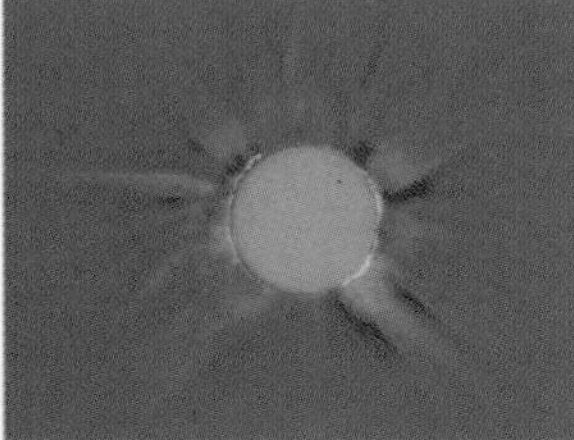

Plate 8

Fig. 1 (top left)
Fig. 2 (top right)
Fig. 3 (left)
Fig. 4 (bottom)

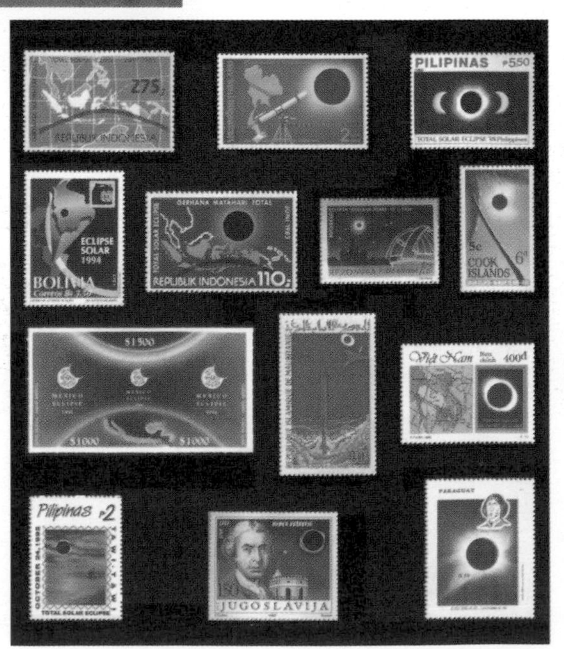

Fig. 4.37. The experiment conducted by the Institut d'Astrophysique de Paris/CNRS aboard Concorde 001, on the eclipse flight of June 30 1973 over Africa. Note the suspension system on the optical-quality window fitted into the roof of the fuselage, allowing the eclipse to be photographed continuously. In the background is Jean Bégot, who carried out the photographic observations during the flight. (British Aircraft Corporation.)

window laminated with Irtan 2. At the Institut d'Astrophysique de Paris, a device was being prepared for photography, to be slung under a 130-mm wide quartz window with a surface accuracy of 1½ wavelengths.

The anticipated duration of totality made possible, for the first time, the study of the complete evolution of small coronal structures, and even a search for pressure waves which might point the way towards an understanding of the origin of heating in the corona. The instrument manned by J. Bégot of the Institut d'Astrophysique de Paris was equipped with three lenses – one of 3 m focal length, which provided the best images taken during totality. The venture was a complete technical success. All the optical components contained within the supersonic laboratory performed well, all the experiments worked, and considerable data were gathered: more than 400 images of the corona, from several cameras, over 74 minutes of totality – far more than was expected initially, and worthy of detailed comment.

These experiments needed both a long period of totality and the longest achievable observation time before second contact and after third contact, in order to prolong as much as possible the passage of the lunar limb across the chromosphere. The planned flight path followed an arc across the eclipse zone,

Fig. 4.38. One of the images of the corona obtained with a 3,000-mm focal length instrument, on 70-mm black-and-white film, during the Concorde 001 eclipse flight of 30 June 1973, when totality of 74 minutes was observed from the aircraft. (IAP/CNRS.)

over Africa, lasting about 1 h 30 m. The demands on the pilot were great: the maximum margin of error allowed for reaching the theoretical rendezvous point of second contact was 15 seconds. Concorde took off at 10 h 08 m from Las Palmas in the Canaries, and flew into the great arc traced out by the shadow. A. Turcat, the well-known test pilot, was at the controls, and he reached the rendezvous point – after 45 minutes in the air – one second early! Wind and temperature conditions were not of the best, and speed was reduced to mach 2.05, assuring 74 minutes of totality instead of the 80 minutes anticipated.

During the flight, altitudes reached varied between 16,200 m and 17,600 m. One of the problems encountered was, naturally, image quality and stability. We were surprised at the good quality of the seeing during short exposures, which compensated for their brevity due to air eddy disturbances; but analysis of images showing the edge of the Moon unambiguously recorded the presence of optical aberrations like distortion due to the thin atmosphere outside the aircraft, since the line of sight had inevitably to pass through a zone, next to the fuselage, filled with turbulent airstreams compressed by the shock wave. These effects militated against very small-scale study of coronal images, causing some disappointment, and it was at first decided not to publish those results; but some compensation came in the form of a great number of ground-based observations of the same eclipse. This type of experiment was never repeated, given its cost and necessarily limited scientific

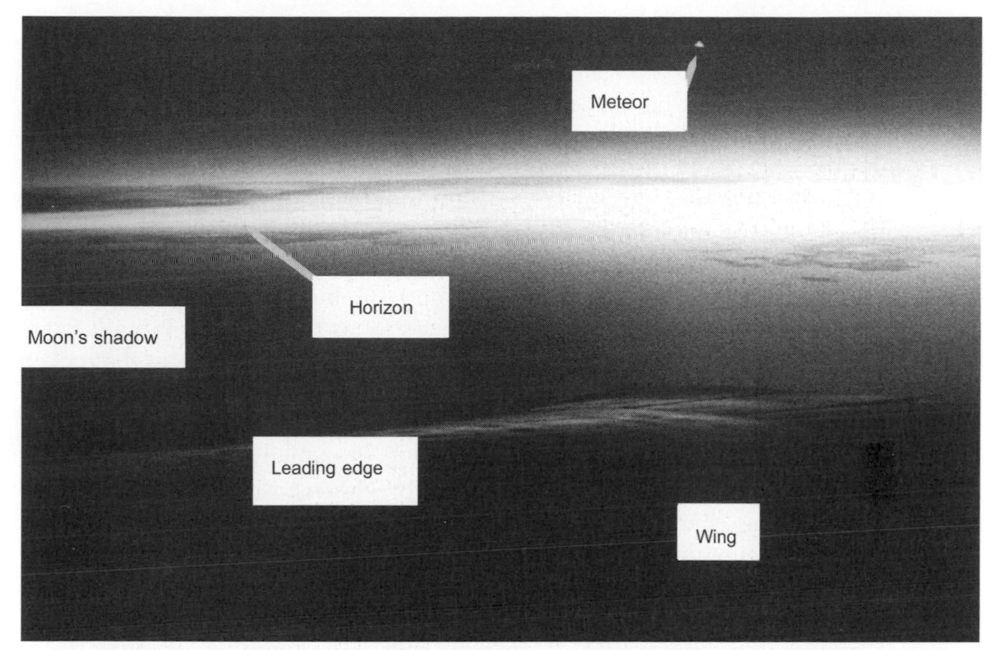

Fig. 4.39. Photograph taken though a window of Concorde 001 during the total eclipse of 30 June 1973, over Chad. The wing of the aircraft can be seen in the Moon's shadow, together with the edge of the shadow and a cloud produced by the passage of a meteor from the β Taurid stream. (J. Bégot, personal collection.)

feedback, but this should not of course detract from the value of airborne eclipse observations. Because of improved modern detectors, even subsonic flights nowadays offer precious extra minutes of coronal study, during the course of which such critical measurements may once again be attempted.

Some of the Concorde images were excellent, showing, for example, coronal polar plumes varying little over the 74 minutes of observation, confirming their magnetic nature. One unrelated photograph, however, was shown around the world, and is still a talking point: on one of the few wide-angle photographs taken by J. Bégot through an auxiliary window just after he had finished his work on the corona, there is an extended object in the sky. What scientific value there is in this image is a matter of debate, but its interpretation has been hotly discussed: UFO believers have seen in it evidence to support their claims. A special report by the CNRS decided that this was undoubtedly a cloud produced in the upper atmosphere, at about 50 km altitude, by the explosion of a meteor from the β Taurid stream, which reaches peak activity on that very day, 30 June, each year. Daylight meteors are well known, due to the radar signatures produced by the large clouds of plasma, or ionised gases, which they create in the upper atmosphere.

OTHER OBSERVING PLATFORMS

Large numbers of rockets have been launched carrying payloads designed for the study of the corona in UV, extreme UV and even X-rays, and many original results have been obtained. Many of these launchings, from established ranges outside the total eclipse zone, have been carried out with the aim of securing data on the corona to complement ground observations, ensuring near-simultaneity. Such was the case with the remarkable rocket-borne experiment mounted by the Center for Astrophysics/Smithsonian Institution, with which L. Golub and colleagues obtained excellent X-ray images of the corona at the same time as we were observing it from Mauna Kea on 11 July 1991. In a certain number of instances, the nose of the rocket remained in the Moon's shadow during the eclipse, allowing some unique measurements. The most memorable experiment was prepared by the Culham Laboratory in the UK, which became noted for its space initiatives involving the extreme UV domain in the late 1960s. The eclipse of 7 March 1970 provided a 'deep' spectrum, with the assistance of American laboratories, since the launch took place from Virginia, where totality was visible; for the first time, an entire spectral image of the corona in the Lyman-α line was obtained. A new property of the corona, involving great potential, had been discovered by Alan Gabriel. Even at such extreme temperatures, some neutral hydrogen could exist, if only in tiny quantities, but it would be enough to permit measurement in the corona of the profile of this line of resonance, present throughout the Universe. This is a property widely utilised nowadays in space experiments involving the study of the solar wind, since the Lyman-α line is a reliable indicator of plasma velocities.

Another experiment deserves to be mentioned, as one of the rare examples of the use of balloons launched from the zone of totality into the stratosphere. This was undertaken by Japanese scientists for the Java eclipse of 1983, with the aim of measuring infrared emissions from the dust corona. A technical success, the experiment produced results which were announced with some *éclat* in specialised publications, and in *Nature*; but their interpretation is debatable. In the light of more convincing ground-based measurements from the eclipse of 1991, using infrared imaging equipment, nobody now accepts the existence of intense thermal rings around the Sun. Infrared studies now concentrate on coronal emission lines, which could in the future – and involving the expected Zeeman effect – lead to direct measurements of the magnitude of the magnetic field in the corona.

Eclipses, including total eclipses, can be observed from almost any point on planet Earth, and also from a point in its vicinity; for example, from a space station or a satellite. Such was the case when G. Nikolsky tried, for the first time, to take advantage of the docking of the orbiting craft Apollo and Soyuz. During the occultation of the Sun by Apollo, photographs were taken with a 70-mm camera through a window in the front of Soyuz. These images revealed magnificent plumes emitted by the small gas-jet stabilisers on the approaching Apollo craft. Eclipse observations cannot be achieved from the orbiting solar laboratory SOHO, as the satellite is too far from the Moon's umbral cone. SOHO is at the Lagrangian point L1 – between the Earth–Moon system and the Sun, and 1.5 million km from the

Earth. Also, observing a total eclipse of the Sun by the Moon from orbit is of limited interest, as occultations from spacecraft are brief. Space vehicles move very fast (typically at about 10 km s^{-1}), and pass rapidly through the Moon's shadow. Partial phases, though, last longer, as the penumbral shadow is more extensive, and many observations of partial eclipses have been carried out from satellites.

Among these interesting spaceborne observations, let us focus on those made from the manned orbital space station Skylab during the total eclipse of 30 June 1973. The Sun was observed from Skylab on every orbit of the Earth, at optical, UV and X-ray wavelengths. The external coronagraph was able to produce a sort of artificial eclipse, revealing the corona; and due to its field width of several solar radii, it was able to watch the Moon occult the corona. This observation proved very useful in measuring the effects of parasitic light, scattered by the calibration instruments incorporated in the coronagraph. The Moon's surface will, of course, appear very dark, illuminated only by light from the surface of the Earth (earthshine). So, very accurate corrected photometry of the corona from space was possible, with some 40,000 images collected during a period of 18 months.

The Japanese satellite Yohkoh, operating since 1991, has already sent back millions of remarkable images of the corona in soft X-rays, at wavelengths around 2 nm. Yohkoh describes a circular orbit around the Earth in 97 minutes, and can, therefore, pass through the penumbra of the Moon several times. Its images have proved fundamental in the evaluation of parasitic scattering effects from instruments, and of cosmic 'noise' in detectors. This is not a simple procedure, even though it is obvious that no X-rays will be observed coming from the lunar disk which is occulting the Sun's disk. Yohkoh has observed several eclipses, and images gathered at these times have also enabled us to see how the performance of the CCD camera used for X-ray studies, and of telescope filters, has improved with time.

So, eclipses are observed even from orbiting platforms, with results of interest to science, especially when those results are complemented by near-simultaneous observations from the ground.

The following summarise historical progress towards space coronagraphs using an *artificial* total eclipse to 'see' the corona.

1930 B. Lyot invents the internally occulted apodisation coronagraph, for use at high-altitude sites. By 1928 he has used external occultation to observe the planet Mercury.

1948 J. Evans (USA) and scientists in Europe develop the externally occulted aureola photometer.

1960 G. Newkirk and J. Eddy carry out the first balloon-mounted photometric observation of the solar aureola with a single external occultation disk, at an altitude of 25 km in the stratosphere.

1963 Using a rocket probe, a team from the Naval Research Laboratory (R. Tousey, M. Koomen and colleagues) obtain the first image of the outer corona of the uneclipsed Sun, using a miniature photographic coronagraph incorporating one small external occultation disk, known as a serrated disk. From the Algerian desert, a French test launch of a rocket

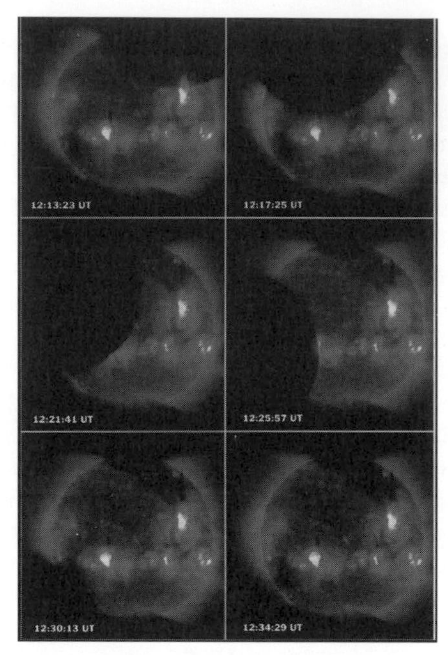

Fig. 4.40. The eclipse and the inner corona seen through the SXT grazing incidence telescope at soft X-ray wavelengths aboard the satellite Yohkoh, during the eclipse of 3 November 1994. Because of the dark lunar disk, the eclipse images allow Yohkoh's detectors to be calibrated, together with levels of parasitic light in the telescope. (Japanese Institute of Space and Astronautical Science/Lockheed Palo Alto Research Laboratory.)

carrying a UV coronagraph by R. Bonnet and G. Courtès, employing a highly polished primary mirror.

1964 Coronascope II, a balloon experiment by D.E. Bohlin and G. Newkirk of the High Altitude Observatory, Boulder, Colorado. Many images of the mid-corona are obtained with a three-disc occultation system.

1965 First satellite flight of a coronagraph on OSO-2 by the Naval Research Laboratory, using a compact three-disc occultation system.

1967 R.M. MacQueen uses a three-blade occultation system and sweeping photometer to record the infrared dust corona from a balloon. Another such experiment is carried out with a mirror coronagraph developed by R.N. Smartt.

1967–68 Balloon experiments carried out by French team of A. Dollfus, B. Fort and colleagues, to photograph the mid-corona with a near-infrared film and an external occultation coronagraph with a serrated disk. First time-sequence of coronal streamers.

1963–72 Rockets launched with two coronagraphs side by side to obviate the effect of shadow produced by the support of the external occultation disk.

1971–74 First daily recording of the corona with a coronagraph equipped with SEC Vidicon camera on OSO-7. First observation of a coronal mass ejection on the occasion of a major solar eruption.

1973–74 During the Skylab mission, the three-disc externally-occultated coronagraph of the Boulder High Altitude Observatory, on the Apollo Telescope Mount, obtains 35,000 high-quality images. Other instruments, operating in the extreme UV and especially in X-ray wavelengths, reveal a completely new picture of the corona. Scientists learn that there is a powerful X-ray corona around the Sun and many other stars.

1976 First flight of a Soviet rocket probe with an external coronagraph built by scientists at the University of Kiev, assisted by IKI Moscow.

1978 Introduction of a new type of externally-occultated coronagraph, with a wider field, in advance of space missions out of the ecliptic. Postponed by NASA.

1979–85 Long series of coronal observations carried out by the Naval Research Laboratory, from the Navy's military satellite P78-1, in polar orbit. Discovery of Sun-grazing comets, and systematic study of coronal mass ejections. The satellite is brought down by a USAF missile on the orders of President Reagan, who is at the time embarking upon his 'Star Wars' programme, the Strategic Defense Initiative (SDI).

1980 Observations begin using a new generation of coronagraphs with external occultation, on board the Solar Max satellite. Unconvincing colorimetric and polarimetric measurements. Many high-quality images of mass ejections. Instrument repaired by P. Nelson and other astronauts from the Space Shuttle *Challenger*.

1990 Launch of spaceprobe Ulysses, out of the ecliptic, to study the solar wind. No imager or coronagraph.

1991 Launch of Yohkoh, specialising in high-energy studies of the Sun. Millions of X-ray images are returned, using a grazing-incidence telescope and CCD, revealing the permanently dynamic nature of the corona. No white-light coronagraph.

1995 SOHO (SOlar and Heliospheric Observatory) launched towards the Earth–Sun Lagrangian point, carrying several coronagraphs with CCD imaging: *a.* monochromatic internally-occultated coronagraph with highly polished objective mirror; *b.* polarimetric white-light coronagraph with 'short' conical external occulter, of excellent photometric quality; *c.* very wide field white-light coronagraph, with three-disc occulter; *d.* Extreme Ultraviolet Imaging Telescope (EIT).

1996 Numerous coronagraph observations by SOHO. Discovery of halo mass ejections, plasmoid flows, rapid MHD waves, ephemeral EUV streams, and so on.

1997 Launch of Transition Region and Coronal Explorer (TRACE).

FILMING ECLIPSES

Cinematographic techniques are of great interest to researchers of eclipses, and to science in general, in both imaging and spectroscopic aspects. With solar eclipses so brief and sudden in their nature, images in rapid succession, on moving film, can be used to study the evolution of events such as the successive occultations of the Sun's disk, the chromosphere and the corona, by the Moon's edge at second and third contacts. Moreover, filming totality at high spatial resolution enables investigation of phenomena such as coronal plasmoids.

The following details the main stages in the use of cinematographic techniques:

1874 J. Janssen is the first to use a rapid-photography technique, with the aid of a rotating element developed in order to record the passage of Venus across the Sun's limb.

1900 First attempt at filming the UV spectrum during a total eclipse, with a wide-plate coronagraph (500 images at 4 per second), on 28 May at Argamasilla in Spain, by Deslandres and Fallot.

1912 Many films made of the European eclipse by De La Baume Pluvinel (Saint-Germain), Carvallo (Trappes), Da Costa-Lobo (Portugal), Vlès (Spain) and R.P. Willaert (Namur). First trichrome 35-mm colour film of the eclipse by Gaumont.

1935 First films of prominences made by Lyot with his coronagraph at Pic-du-Midi Observatory. With J. Leclerc, he produces the film *Flames of the Sun*, showing mighty eruptive prominences, with explosions, ejections and condensations of matter.

1936 Excellent films showing the contacts and totality, at 2 frames per second, from the expedition of the Société Astronomique de France to the USSR eclipse of 19 June. Their success was repeated during the eclipse of 9 July 1945.

1952 Colour films (Gevacolor and Eastmancolor) by several teams in the Sudan, 25 February, reveal pink prominences against the white corona. Many such films have since been made.

1991 Colour video films with large-format cameras (1,200 lines) obtained by the Japanese NHK Television team in Hawaii.

Nowadays, very sensitive high-resolution CCD video cameras provide us with excellent films of the contacts, and of totality.

5

Observing total eclipses of the Sun

Between first contact and last contact of an eclipse of the Sun, many things happen, often in rapid succession. What are the phenomena to be observed? How does one observe a total eclipse of the Sun? Here is some advice for the 'eclipse chaser'.

PREPARATION AND TRAVEL

During an eclipse of the Sun, the Moon's shadow falls upon only a limited part of the Earth's surface. This means that the probability of a total eclipse being visible from any place on the Earth is extremely low, so it is usually necessary to travel to the zone of totality.

The choice of an observing site depends on three factors: ease of access to the country from which the eclipse is to be seen; duration of the eclipse; and weather prospects for the chosen site on the due date. Finally, in the case of far-flung locations, the attractions of the country to be visited are an additional consideration.

Accessibility of the observing site

A question often asked by those who are new to total eclipses is: why should I have to travel to see them? Celestial mechanics provides the answer. The distances from the Sun to the Moon and from the Moon to the Earth are respectively 149,600,000 km and 384,000 km, the first distance about 400 times greater than the second. The respective diameters of the Sun and the Moon are 1,391,000 km and 3,476 km – again a ratio of 400:1. It is a remarkable coincidence that the Moon is 400 times smaller than the Sun, but 400 times nearer. Therefore, the two bodies appear to us to be about the same size in the sky, and the Moon, in its orbit around the Earth, can pass in front of, and hide, the Sun. Since the Sun shines upon the Moon, the latter casts into space a conical umbra and penumbra. The length of the umbral cone is 350,000 km ± 25,000 km. When the cones are intercepted by certain parts of the Earth's surface, observers in the regions involved will see an eclipse of the Sun – total in the case of those areas swept by the umbral cone, and partial in the penumbral cone. The curved paths of these lunar shadows may be thousands of kilometres long.

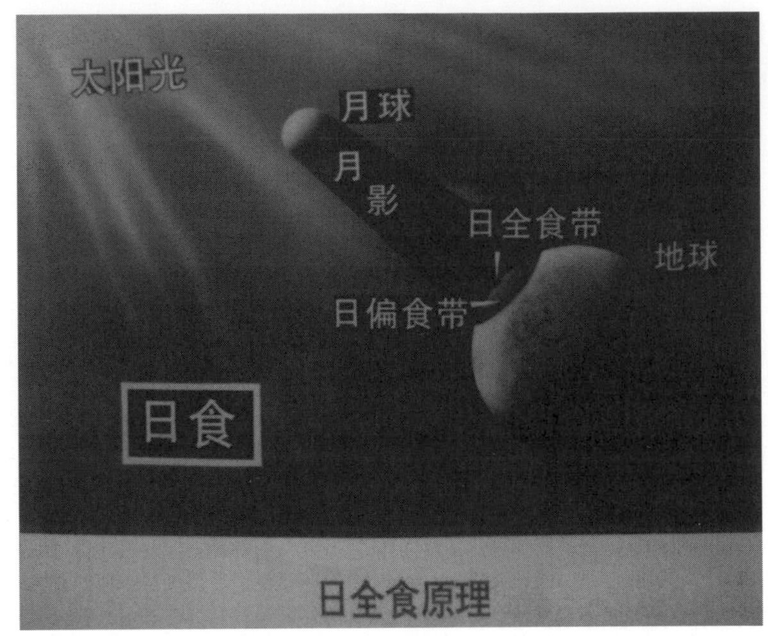

Fig. 5.1. Chinese *dazibao* poster, photographed in Peking, illustrating the phenomena of the umbra and penumbra during an eclipse of the Sun. (P. Guillermier.)

The Moon's shadow sweeps out a band across the Earth's surface up to 269 km across.

The choice of an eclipse observing site is decided by tracing the centre line of the eclipse on a map of the world, and listing the countries through which it passes. From this list, the eclipse chaser may select certain easily accessible and inexpensive countries, which offer a reasonable guarantee of safety. Countries should be able to offer reliable accommodation and transport, so that the observer can be well rested and able to arrive at the chosen site in good time. The beauty of the country visited may be a criterion, if tourism and astronomy are to be combined.

Since the eclipse of 11 July 1991, which attracted several hundred thousand observers to Hawaii, Baja California and neighbouring regions of Mexico, many American travel agencies have been offering eclipse trips, and cruises on well-appointed liners are organised. This is an elegant way to avoid problems with crossing frontiers, and very apt in that eclipses tend to occur mostly over seas and oceans. It should, however, be borne in mind that the deck of a ship is in constant motion, and long-exposure photographs are out of the question.

(A list of travel organisations is included in the addresses section towards the end of this book.)

Duration of the event

The length of totality is an important consideration in the choice of site. As has already been mentioned, the theoretical maximum is 7 m 58 s of totality, at the equator. Precious seconds of totality can be gained by positioning oneself exactly on the centre line, allowing time to complete detailed observational and photographic programmes. The duration of an eclipse for a given site is normally determined by referring to ephemerides, but this is more often done nowadays by using computer software. Reference can be made to the *Canon der Finsternisse* by Theodor von Oppolzer, which lists eclipses from 1207 BC to 2161 AD, or the *Canon of Eclipses* by F. Espenak, covering the period 1986–2035. (Further details, together with software titles, are included in the addresses section towards the end of this book.)

Weather prospects

Atmospheric conditions are the bane of astronomers, and especially of eclipse watchers, who may see months of preparation, and a journey to the other side of the world, wasted by adverse weather. To fully sympathise with that dark moment in an astronomer's life when an eclipse is clouded out, ruining months of preparation, it is necessary to have lived through it oneself.

This kind of disappointment can be better appreciated by consulting the works of Jean-Baptiste Le Gentil de la Galaisière, an astronomer at the Paris Observatory, who was sent by the Académie des Sciences to Pondicherry to observe the transit of Venus across the face of the Sun. His two volumes bear the title *Voyage dans les mers d'Inde (1760–1771), fait par ordre du roi, à l'occasion du passage de Vénus sur le disque du Soleil, le 6 juin 1761 et le 3 du même mois 1769*. The unfortunate astronomer was sent to measure the disk of Venus, standing out as a black dot against the Sun's disk, with a view to estimating the dimensions of the solar system. Transits of Venus occur in pairs, eight years apart, at intervals of approximately 110 years. On 6 June 1761, the war between France and England prevented Le Gentil from observing the transit from Pondicherry. For the transit of 3 June 1769, Le Gentil was undecided whether to go to Pondicherry or Manila, and decided on Pondicherry as a more politically stable place. But, as he sadly recounted: 'Such is the fate that often befalls astronomers. I had travelled more than ten thousand leagues. I had, it seems, crossed such a vast expanse of ocean merely to be a spectator of the fateful cloud which passed before the Sun at the very moment of my observation. Snatched away were the fruits of my labours and of the weary hours. And, while the sky used me thus at Pondicherry, it was at its clearest, as I later learned, in Manila'.

Today, we have some modern tools to help us evaluate weather prospects on the day of the eclipse. Information on mean cloud cover over eight years, derived from international weather surveillance satellite data, is available to those seeking a clear sky. These data, together with the frequency of highly reflective clouds (detectable by satellites in visible light and in infrared), and mean values for hours of sunshine per day, may be found in NASA publications, or on the Internet. (References to these publications and Internet sites are included in the addresses section towards the end of this book.)

INDEPENDENT 35p

news

Solar eclipse: Asians gather to see Moon obstruct morning light, but India's superstitious stay inside to ward off evil spirits

Millions wonder at darkness in the heavens

Estrella DE ARICA

Eclipse

Más de 11 mil fueron a Putre

Gráficas espectaculares

Emoción por el eclipse total

Una niña con lesión ocular

LA TERCERA

El día se hizo noche en Putre

Impactante eclipse solar

Fig. 5.2. Newspaper articles featuring precautions necessary when observing eclipses. Useful information may be gleaned from local newspapers, especially lists of towns on the centre line, local times and durations of totality for these sites, and weather conditions expected. Such data may complement information gathered before the trip.

EQUIPMENT

BEWARE! never look directly at the Sun without protection!

During the partial stages of an eclipse, it is important never to look at the Sun without protecting your eyes. Even when partly hidden, the Sun is so bright that it may cause serious damage to the retina, and there is the risk of being blinded. The danger is heightened by the fact that the retina does not feel pain, and its destroyed cells will never recover.

To observe in safety, therefore, it is necessary to use filters, the best of which are made of welder's glass. This may be bought cheaply, and No. 14 will ensure safe observation of the partial stages. Glasses specially made for eclipse viewing can also be used, although mylar filters are best avoided, but if used must be multi-layered and in perfect condition, without wrinkles or scratches. 'Blackened' colour film should never be used, even when 'stacked'.

Eclipse observers' equipment should satisfy various criteria. It should be:

- compact, so that it can travel as hand luggage on aircraft. Travelling in a cargo hold is not recommended for fragile optics; customs certificates should be to hand to avoid delays at frontiers;

Fig. 5.3. Amateur astronomer Roland Caron – a very active member of the Société Astronomique de France (SAF) during the 1980s – prepares one of his experiments in his hotel room before the eclipse of 22/23 November 1984 (the central line crossed the International Date Line). The SAF team, led by Christian Nitschelm, successfully observed from the afterdeck of a naval ship, the *Jacques Cartier*, off Nouméa (New Caledonia). (SAF.)

- light, as you may have to carry it on more than one stage of the journey;
- robust, and shock-resistant; bubble-wrap or similar packaging should be used to protect instruments, cushion jolts and vibrations, and prevent misalignment of optics;
- reliable. What a pity it would be to miss the spectacle as a result of some rough-and-ready construction work, and spend the duration of totality repairing your instrument. Make sure that the systems you use are tried and tested in advance.

Equipment may be limited merely to a pair of suitable binoculars. This lightweight instrument, magnifying usually up to x10, will allow a view of the whole phenomenon, and the diameter of its objectives can enhance the brightness of the eclipse. A small 60-mm refractor, a 100-mm Schmidt–Cassegrain reflector or a 90-mm Maksutov–Cassegrain reflector may be mounted on a photographic tripod. With these instruments, detailed study of prominences, the chromosphere and the corona will be possible, using a previously selected low-power eyepiece. A pair of binoculars on a tripod allows the corona to be studied with both eyes.

Partial stages must be observed with the same eye protection as if the complete solar disk were being viewed. During the last moments of visibility of the photosphere, the Sun is still bright enough to cause eye damage if observed directly.

Fig. 5.4. As with the non-eclipsed Sun, partial stages must be observed using eye protection of very dense filters. The Schmidt–Cassegrain telescope shown has a full-aperture filter. The astronomer on the right is wearing mylar glasses, and the other astronomer, behind, has welder's glasses. The temperature during this eclipse in Mongolia on 9 March 1997 was –20° C, which accounts for the clothing worn. (Y. Méziat.)

The best way to appreciate second and third contacts is with direct vision, and optical instruments used must be equipped with filters, or the Sun's image may be projected onto a screen. If no optical instrument is used, welder's goggles, black polyethylene eclipse glasses or mylar glasses should be worn.

It is also possible to pierce a pinhole in a sheet of card, and project the Sun's image onto another piece of card. This *camera obscura* technique was used by French painters in the eighteenth century to create miniatures, but its origin dates back to the English Franciscan monk Roger Bacon (1212–1292). The smaller the hole, the more contrast the image will show; a 1-mm hole provides a good balance between brightness and contrast. The diameter of the Sun's image will be 1% of the distance between the two cards. Another, even more simple, method is to reflect the Sun's image onto a screen from a small mirror. Again, the size of the image will depend on the distance; with the mirror at a distance of 10 m, the image will be 10 cm in diameter.

THE ECLIPSE SCENARIO

A total eclipse of the Sun can be divided into three periods. During the first, the Moon hides part of the Sun; then, for a matter of minutes, comes totality; and finally, the Sun gradually regains its circular appearance. First contact, marking the beginning of the eclipse, is the moment when a tiny portion of the Moon appears in front of the Sun. In reality, a tiny black notch appears. From this moment, the Sun's disk is slowly hidden by the Moon. This is almost entirely due to the Moon's own orbital motion around the Earth. During the first part of the eclipse, and until 50% of the Sun is hidden, there is no noticeable difference, to the naked eye, in light levels. Beyond this, the diminution in brightness becomes detectable, if difficult to appreciate because of the slow progress of the phenomenon. As the Sun shrinks, shadows appear more contrasted, and sunlight reflected from water seems to sparkle more than usual. Temperatures fall noticeably after 50% of the Sun is hidden. If the sky is clear, it is of an intense, deep blue, as a result of the decrease in brightness.

- One or two minutes before the onset of totality, shadow bands appear. These are dark striations which appear to move across light-coloured surfaces.
- 40 to 20 seconds before totality, the last light from the thin solar crescent is blocked by elevated features on the Moon, though the light still passes through its valleys. Now, the Moon's edge resembles a string of bright diamonds, a phenomenon explained by the English astronomer Francis Baily, and called *Baily's Beads*.
- 20 to 10 seconds before totality, from the western horizon, a wall of darkness rushes in at supersonic speed, rearing up into the sky and engulfing the observing site. Being enveloped by the shadow of the Moon is one of the most spectacular and impressive moments of a total eclipse.
- 5 to 3 seconds before darkness, the last vestiges of brightness give way to a ring of phantasmal light, as the pink chromosphere and its prominences begin to appear. At the moment of second contact, the Moon covers the last part of the photosphere. Totality has begun, and the eclipse can be observed safely without filters.
- 3 to 5 seconds after second contact, the pink ring of the chromosphere develops around the Moon. Prominences – moving condensations of 'cool' coronal gas, supported by the magnetic field – are now seen. These red tongues are often visible shortly before and after totality. At mid-totality during the longest eclipses (5–7 minutes' totality), the lunar disk may be large enough to hide most of the prominences.
- The corona – the outer part of the Sun's atmosphere – is seen after second contact. As soon as the eye is adapted to the darkness of totality, the corona appears, looking like a flower with white petals surrounding a black centre. The white-light corona is composed of highly ionised gases (plasma), disturbed by magnetic fields. The appearance of the corona varies from one eclipse to the next. At solar minimum the corona appears symmetrical in relation to the Sun's polar axis, but near the time of maximum activity in the 11-year solar cycle, the corona may be extremely disturbed.
- At mid-totality, the observing site is at the centre of the Moon's shadow. Illumination around the horizon gives the impression of a 360° dawn breaking. The sky may now be dark enough for planets and stars to magnitude 3 to be seen with the naked eye.

Fig. 5.5. One of the best images of the corona in the red light of FeX at 637.4 nm. Taken on Kodak T-Max 400 with a 150-mm lens of focal length 3 m, exposure 8 s, with an interferential filter of 0.3 nm bandpass. At the top is the eastern limb, with west below. Eclipse of 11 July 1991 from La Paz, Baja California, Mexico. (Joint Kwasan-Hida Observatories team, courtesy H. Kurokawa.)

The end of totality is the moment when the Moon allows the Sun's light to be seen again, and a small portion of the photosphere emerges. This is third contact, and eye protection must again be worn. The first-to-second-contact process now runs backwards, until, at fourth contact, the Sun regains its normal appearance. Here we can quote Francis Baily, a merchant and also an amateur astronomer, who wrote this very faithful description of the feel of the phenomenon: 'I was amazed by the great uproar of applause coming from neighbouring streets, while myself electrified at these phenomena, of the most *brillant*, *splendide* * nature imaginable. In an instant the black body of the Moon was suddenly surrounded by a corona, a kind of glorious necklace. I had anticipated a luminous circle around the Moon during the minutes of totality ... but the most remarkable thing was the appearance of three great *protubérances**, apparently emanating from the lunar circumference, but evidently forming part of the corona.' (* In French in the original.)

AN OBSERVING PROGRAMME FOR THE AMATEUR

During an eclipse, events succeed each other very rapidly; and some events occur simultaneously. The short duration of totality necessitates advance planning of observations. It will simply not be possible during this brief period to look for comets, for example, while hoping also to study prominences or the chromosphere. According to individual preference, an observing plan, timed to the second, should be drawn up. Table 5.1 lists the order of events and phenomena, and will help in planning observations and photography. For each stage, let us examine optimum observational techniques.

Table 5.1. Events and phenomena during a solar eclipse

Events	Solar phenomenon	Ambient phenomenon
First contact	Moon's disk moves onto Sun	
Partial stage	Moon moves across Sun	Light levels fall
		Temperature falls
		Animals behave abnormally
Second contact	Baily's Beads	Shadow bands
Totality begins	Chromosphere	Shadow arrives
	Prominences	Onset of darkness
Totality	Corona	Stars/planets visible
		Comet(s) possibly visible
Third contact		Shadow departs
Partial stage		
Fourth contact	Sun's disk back to normal	
	Moon's disk no longer visible	

FIRST CONTACT

First contact is an emotional moment. After all the preparations, to be in the right place at the right time and to have provided for the moment is a source of great satisfaction.

The Moon meets the Sun

This moment is detectable with the naked eye, protected with welder's glass or mylar. A magnification of x5–40 is needed to see this easily, through an instrument equipped with a filter, or by projection for safe viewing.

The exact time of the event depends on whether calculations made, or ephemerides consulted, are reliable. Remember that ephemerides usually provide times for places where the eclipse is of maximum duration. If the observer is not in this place, corrections will have to be applied. Also, ephemerides are expressed in Universal Time (UT), and it will be necessary to add or subtract the requisite number of hours for the site.

The Moon moves across the Sun

The disappearance of the surface of the Sun behind the Moon's disk may be observed with the same instruments as those normally used for solar observation. Magnifications used range from x1 (naked eye), to a maximum x40, if the Sun is to remain entirely within the field of view of the instrument. Whether observations are carried out with the naked eye or with an optical instrument, it is vitally important to use adequate filtration, or a safe projection method, to avoid running the risk of lasting and serious eye damage.

During the partial stages, shadows on the ground from leaves on trees take on an unusual aspect, as each space between the leaves no longer casts a circular image of the Sun, but a small crescent-shaped one, faithful to the shape of the eclipsed Sun. Also, because of the reduction in the Sun's apparent surface area, reflections on water have more sparkle, and shadows are sharper.

Falling light levels

Any decrease in brightness is not really detectable until 50–60% of the Sun is covered. Visually, however, the impression is hardly more than that of a cloudy sky. It is only when 80–90% of the Sun is lost that light levels become noticeably dimmer than when the sky is overcast, and the colour of the sky looks very different from normal.

It is possible to record this, and to draw a neat curve of brightness variations during the event with a light-meter, such as those found in cameras for determining exposure times and aperture settings.

Fall in temperature

As the surface of the Sun is covered by the lunar disk, energy received will decrease. Temperature is an indicator of this energy, and may be measured during the eclipse.

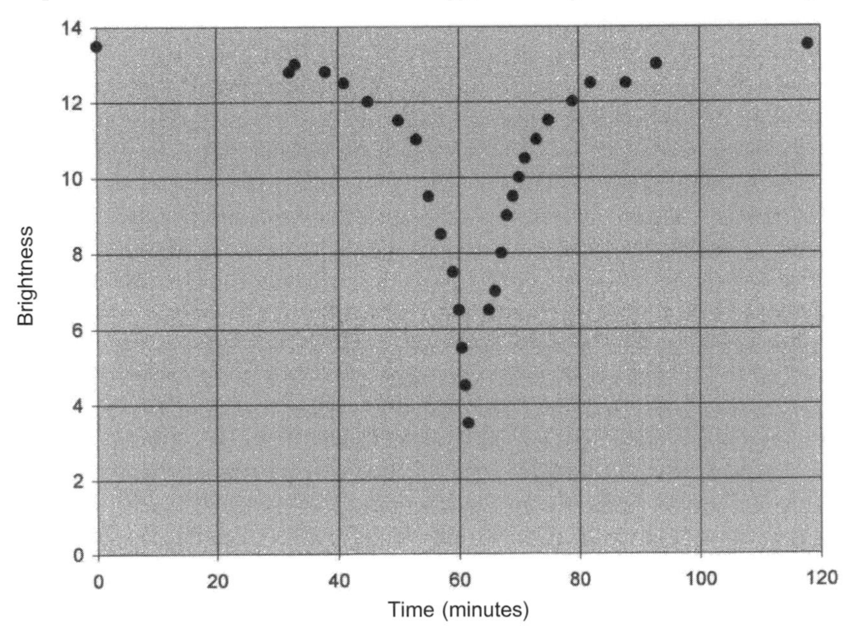

Fig. 5.6. Curve showing brightness variations measured against time, made from readings with a camera light-meter during the eclipse of 3 November 1994. (B. Druel.)

Temperature measurements should be taken in the shade, at least 1 m above ground with a covering of grass. Decreases in temperature, which vary from one eclipse to the next, are related to the duration of totality and to atmospheric conditions. They range from 5° C to 15° C.

Shadow bands

The phenomenon of shadow bands – a succession of light and dark striations – is somewhat random. German astronomer Hermann Goldschmidt was the first to remark upon this complex refraction phenomenon, in 1820. Shadow bands may be observed on light-coloured surfaces; for example, walls or white sheets, or occasionally on the ground, when the 'crescent' of the Sun is very thin, about one minute before and after totality.

Baily's Beads

As we have seen, this phenomenon is due to irregularities along the lunar limb, as valleys allow light to pass through, while high mountains mask the Sun. Baily's Beads are difficult to observe, as this short-lived phenomenon lasts only 5–10 seconds. Their observation needs a magnification of x10–40, and filters will have been attached to the instrument a few seconds before; care will need to be taken when removing filters. The time taken to do this unfortunately reduces the time available for observing Baily's Beads.

SECOND CONTACT: TOTALITY

The Moon now covers the Sun completely. Observations can now proceed without solar filters, which should be rapidly removed from the optical path of the instrument, in order to gain as much observing time as possible.

To adapt the eyes to the darkness, some observers take the precaution – several hours before the eclipse – of blindfolding one eye. At the instant of totality, the eye – its pupil dilated by hours of darkness – will be able to appreciate low-light phenomena. But the 'old wives' tale' current among eclipse observers – that the eyes have time to slowly adapt as the Sun's light gradually fades during the entire partial phase – has no basis in truth.

The shadow arrives

This is the most spectacular and impressive moment of the eclipse. It is easy to imagine the awful fear that must have gripped ancient peoples living through these exceptional events. The shadow travels at between 500 and 2000 m s^{-1}, and the wall of shadow, moving at supersonic speed, envelops the observer in a fraction of a second.

With rare exceptions – for example, near the poles – the shadow moves from west to east. This means that, if a morning eclipse is observed facing the rising Sun, the shadow will arrive from behind. Conversely, for an afternoon eclipse, the shadow will move in from the direction of the Sun.

Fig. 5.7. 180-degree panorama of the Kazakhstan total eclipse, 31 July 1981. Note the edges of the Moon's shadow, delineated by back-scattering in the Earth's atmosphere. On this scale, the Moon's disk is drowned out by the overexposed corona. (Izmiran, Moscow.)

Darkness

Experience, and written accounts, prove that the darkness which reigns during totality is not easy to describe. Several factors combine to distort human perception at this time. These include the suddenness of the drop in light levels (to which humans are unaccustomed, since sunrises and sunsets proceed slowly, due to atmospheric diffraction) and the disappearance of the Sun's light during the day.

The 'darkness' corresponds to a low light level of about 2 lux, which is comparable to the light of the full Moon. It is advisable at this time to have a pocket torch (not too bright) to help with arranging instruments and cameras; and a clip-on or self-supporting lamp is also useful, as it leaves the hands free.

The reddish-orange or yellow glow around the horizon is due to light from the chromosphere and the photosphere outside the shadow zone. The impression is of sunrise or sunset all around. If you are lucky enough to be able to observe the eclipse from near the beginning or end of the shadow track, the shadow will be very elliptical in shape. The 'dawn' glows to the north and south will thus be seen, but the east and west horizons will be plunged into darkness. As for the distribution of brightness and its linear polarisation in the sky during totality, this is a complex matter, varying with the respective positions of the observer and the Moon's shadow on the ground.

Prominences

Prominences often project from the edge of the solar disk, especially when the Sun is very active. The lunar disk therefore briefly mimics the Lyot coronagraph, by

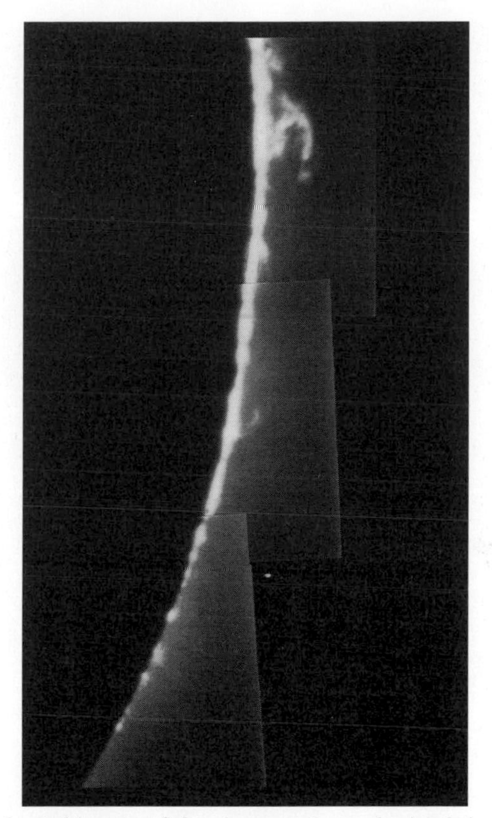

Fig. 5.8. A much enlarged image of the chromosphere obtained through a red filter on Kodachrome 25 ISO, approximately 10 s after first contact (eastern limb), during the eclipse of 3 November 1994. Note the details of the lunar profile, slightly exaggerated by the phenomenon of irradiation by overexposed parts of the chromosphere projecting above the irregularities. (J.-P. Zimmermann and G. Woehrel, and IAP/CNRS.)

masking the photosphere. Observation through instruments with a magnification of x50–100 is spectacular.

The chromosphere

It is possible to observe the delicate pink fringe around the black Moon with a magnification of x7–40, from just after second contact until just before third contact. Spicules and small coronal plasmoid ejections are among the most remarkable events to look for.

Stars and planets appear

During totality, the sky is dark enough for planets and stars down to magnitude 3 to be seen. An eclipse presents an opportunity to observe planets in unfamiliar positions: for example, Mercury and Venus may be admired high up in the sky. Observations may be made with the naked eye.

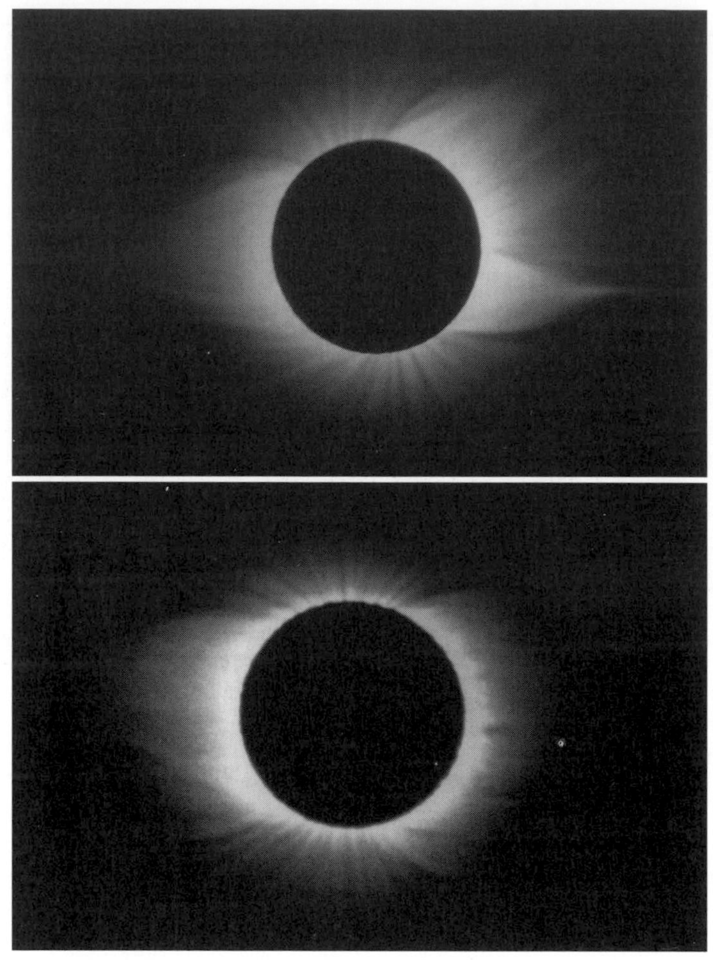

Fig. 5.9. For the eclipses of 1994 and 1995, the Sun was in the minimum phase of its activity. The corona was therefore symmetrical in relation to the polar axis, with large streamers near the equator, and plumes near the poles. The photographs were obtained with a parafocal mask which eliminated overexposure of the inner corona. (F. Diego, University College London.)

The solar corona

This special phenomenon can be observed only during a total eclipse of the Sun. It is not seen from the ground at other times, as its brightness is less than that of the daytime sky. The corona sometimes extends out to ten solar diameters, and may therefore be observed with the naked eye, or with wide-field instruments with a magnification of x7–10 at most.

The shape of the corona depends largely on solar activity. Around solar minimum it is often symmetrical in relation to the Sun's polar axis. Near the poles emerge solar plumes, whilst long streamers reach into space from around the equator. At solar

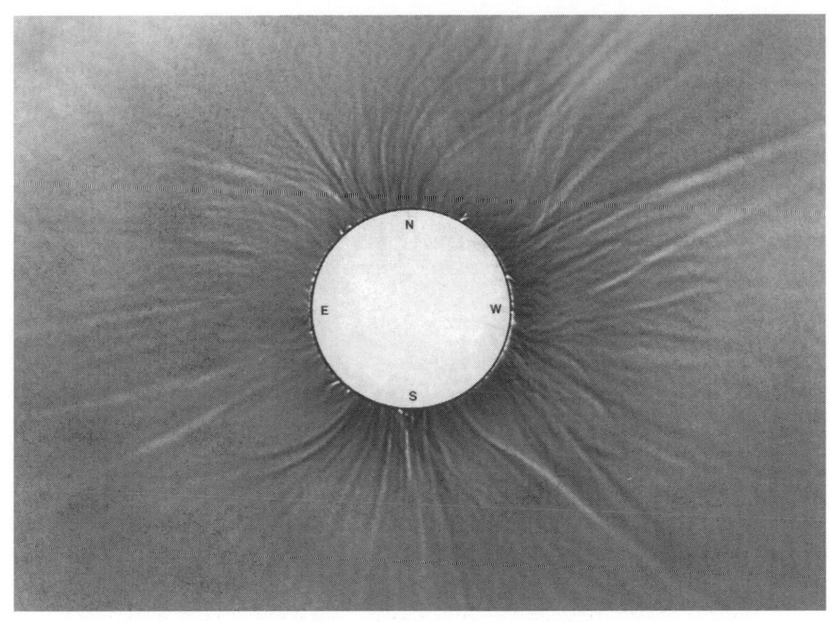

Fig. 5.10. In this picture, the structure of the Sun's corona is brought out by a process of reinforcement of transverse gradients, revealing streamers above the north and south poles. The eye is still the only detector capable of simultaneously appreciating fine, faint details in the corona, and brilliant prominences. The original observation was carried out by the team from the Institut d'Astrophysique de Paris, in Kazakhstan, on 31 July 1981. (S.K., J. Fagot and C. Lebecq, IAP/CNRS.)

maximum, the corona does not show this symmetry. Huge streamers are present all around the Sun, and large mass ejections may often be seen.

Possible comets

Comets may be sought in a region 10–60° from the Sun. Wide-field binoculars, with good light grasp, are necessary.

During the eclipse of 9 March 1997, in Mongolia and eastern Siberia, comet Hale–Bopp C/1995 O1 could be seen with the naked eye, 46° from the Sun, during totality by some observers, through snow-laden clouds. At the time its magnitude was about 0. Comets seen during eclipses of the Sun include one on 17 May 1882 from Egypt, comet Rondanina–Bester on 20 May 1947 from Brazil, and another on 1 November 1948, from Kenya.

Sun-grazing comets – now well known from observations by spaceborne coronagraphs with external occultation – have never been photographed during a total eclipse, a fact which reflects their rarity.

Abnormal animal behaviour

It is very interesting to study the behaviour of animals, and especially birds, which tend to be present. During the disappearance of sunlight, they tend to be very nervous. Birds may panic. Silence reigns, as birds stop singing, during the dark minutes of totality. A simple way to preserve and recall the ambient sounds as totality arrives is by means of a tape recorder.

Many eclipse observers report that, a few moments before totality, cows often move homewards, thinking that sunset has arrived.

THIRD CONTACT

A blinding flash of light signals the end of totality. The photosphere reappears. After this moment, observers receive the impression that darkness has lasted only a few seconds. They try to fix in their memory the spectacle which they have just witnessed, and there is general rejoicing.

The shadow departs

Even though the umbral cone is moving at the same speed as it was during the passage from light to darkness, the onset of light is very sudden. This is rarely noticed, since the observer is often taken by surprise by the return of this avalanche of light.

Reappearance of a fraction of the solar disk

Now is the time to reattach adequate filters to instruments, or to turn back to projection. The same rules apply for the observation of this phase as for the time before totality.

FOURTH CONTACT

The Moon moves off, and the Sun regains its circular form.

Disappearance of the Moon's disk

This is the final moment of the eclipse. It has no particular interest for the observer.

Understanding the nature and sequence of these events implies that, from your very first eclipse, you will have planned your observations carefully. As the eclipse proceeds, you can turn your attention to those atmospheric and astronomical phenomena which offer the most interest, as they occur. However, if you are a first-time observer, one or two events will hold your attention and you will certainly not have time to observe everything described above. So your one aim will be to see the next eclipse of the Sun, to photograph its magnificence, and to take away memories of rare moments.

AMATEUR OBSERVATIONS FROM AN AIRCRAFT

The following account of the total eclipse of 1984, seen from an aircraft flying at 3,000 m, comes from teacher Jean Messud, of Nouméa, New Caledonia. The total eclipse was not visible from the island, but could be seen at some considerable distance from its coast, over the ocean. Messud collaborated with the team from the Société Astronomique de France for this event.

His article, written for a local newspaper, appeared the next day, and is worthy of reproduction here:

> Magenta Airport, Nouméa, 8h 30m, six people aboard a modified Britten at the end of the runway. Engines roar, and we are rolling; a superb take-off. Our destination, a point out over the ocean, some 180 km from Nouméa; our mission, to observe the total eclipse of the Sun. We climbed rapidly, amid intense noise. The partial stage of the eclipse was already under way, and the disk of the Sun had been eaten into. The altitude at which we were flying, and the failing sunshine, caused the temperature within the 'plane to drop rapidly. We were soon wearing the woollen clothes which we had been careful to bring along.
>
> Apart from the pilot, there were five very excited people on board. Some of the seats had been removed to create some much-needed space, and a rear door had been taken away for ease of observation and photography. After a while, we noticed some low cloud, which grew more and more dense. Would our colleagues on board ships be able to see anything?
>
> The eclipse proceeded rapidly. Through welder's glass, we watched the progress of the Moon across the face of the Sun. Colours were already changing: everyone had a very leaden appearance, the reflections of the Sun dimmed, and contrasts were erased.
>
> Darkness soon fell, and, through the filter, only a very narrow crescent of sunlight remained. Zero minus 3 minutes. One observer set himself up beside the pilot, to improve his chances of obtaining good photographs. We went over our schedule, everyone knowing exactly what to do at the appropriate time. Watch synchronised, programme checked, instruments checked for the last time. Tension mounts. Countdown. A few seconds more ...
>
> Our first big surprise: still about ten seconds to totality, but the corona is already there, pale, faint, but IT'S THERE! Venus and Jupiter can be seen, and three or four bright stars. Then, totality. That rarest of spectacles, spell-binding, sublime: near-darkness has fallen.
>
> Seen through the windows of the aircraft, the corona is resplendent, much brighter than we had expected. Coronal streamers could be seen with the naked eye out to two or three solar diameters from the edge of the Moon. Around the Moon's dark disk, the chromosphere draws a thin, bright pink fringe. We can make out hardly any prominences: they must be small. Others did see them, small indeed and bright red. The corona (the Sun's atmosphere) is striking, of a bright whiteness somewhere between milk and pearl.

Picture the scene: the black disk of the Moon, with its fine pink border, all contained within a pearly ellipse of shining streamers of different lengths. Lay this image upon a deep navy-blue sky, and you will have a good idea of what we saw.

Although the sky was dark enough for stars and planets to be seen, around the horizon it was still light, and a strip of sky several degrees high shone with a glow sometimes yellow, sometimes orange, with occasional brown gleams.

Excitement reigned aboard the 'plane. Everyone was very busy and self-absorbed, but all felt the passage of a unique moment. Great activity: the time elapsed is shouted over the noise of the engines, cameras click away, and change hands for resetting. And all in the dark.

Already, the corona is fading, and a growing light surrounds the Sun. A few last photos, and the phenomenon is ended. One or two seconds more, and it's really over. After some comments on eclipses we have seen, and comparisons with this one near Nouméa, we take one last look at the horizon, clearly seeing the Moon's shadow slip away towards other lands. As the sky brightens, the 'plane makes its turn.

It's all over.

A fine description of a modern amateur observation. Some of the photos taken by René Verseau from the aircraft were used for absolute measurements of the intensity of the corona (*L'Astronomie*, 1986, vol. 100).

6

Photographing eclipses of the Sun and Moon

A successful eclipse photograph – a souvenir of the beauty and emotions of the event – is an easy undertaking. It is much harder, however, to capture an image of the fine structures of the solar corona, and to succeed in polarimetry and spectroscopy.

PHOTOGRAPHING SOLAR ECLIPSES

What camera should you choose to take your own souvenir photograph of an eclipse of the Sun? Some phenomena need only the most basic equipment, while others require something more sophisticated.

Experienced photographers – with their refractors or reflectors of focal length 1,000 mm or more – should know that any high-resolution images they may obtain, preferably in colour, during totality, could be of use to the scientific community. Photographs taken as a sequence are particularly valuable, as they can show variations in small structures. (Do not hesitate to send copies of your photographs to: in the UK, British Astronomical Association, Solar Section, Burlington House, Piccadilly, London W1V 9AG; in France, Institut d'Astrophysique de Paris, 98 *bis* boulevard Arago, 75014 Paris.) The sequence of events of a solar eclipse, and equipment needed for successful photography, are discussed below.

Preparing for the eclipse

Preparation will be necessary, given the short duration of totality and the fact that you will be suddenly working in near-darkness. The limited time available during a total eclipse means that the sequence of photographs must be planned with utmost care, and equipment must be tested to ensure faultless performance on the day.

Handling

It should be borne in mind that it will be nearly dark during totality, and cameras will therefore have to be operated in low-light conditions. All possible settings will have to be anticipated. For example, a few tens of minutes before totality, when daylight allows you to see what you are doing, you can load a new film into the camera, set the focus to infinity (∞), or set it by aiming, with a proper filter, at the solar crescent during the partial stage. The aperture (*f*) setting should be at maximum – for example, 2.8 or 5.6 – and the flash (inapplicable for this kind of photography) disconnected or covered. Do not dazzle the other spectators! A pocket torch, or preferably a free-standing or clip-on type, should be available, for use only if absolutely necessary, since shining a light at the wrong time could dazzle. All you need to touch during totality are the exposure setting and the shutter release. Calculating the degree of darkness in advance is tricky, since the amount of light within the shadow is extremely variable. As a rule of thumb, however, the amount of light prevailing during a total eclipse is about 2 lux. Rehearsals can be carried out with photographic equipment by the light of the full Moon, or for a few minutes one hour after sunset.

Simulation

A simulation of a total eclipse of the Sun can be effected in two ways: in a planetarium, and with a computer.

In a planetarium, it is possible to recreate the phenomenon by feeding the correct time and coordinates (latitude and longitude) into the positioning mechanism of the projectors. This type of simulation will provide some idea, on a largish scale, of the altitude of the Sun at the instant of totality, and the positions of the planets in the sky. During the session, you can calmly decide upon the lenses which will be needed during totality; for, when it comes, there will be no time to change them. Simulating eclipses is not something that a planetarium operator is asked to do every day, and you will have to make contact with him and persuade him to see you out of hours, to plan a simulation. You will need to supply the coordinates of the chosen site, and the estimated time of the event in Universal Time (UT). (A list of planetaria is given in the addresses section towards the end of this book.)

There is a great deal of astronomical software which will display on your monitor screen the Sun, and stars and planets visible for an eclipse of your choice. Type in the data for the observing site, and the date and time (UT) of the event. (A list of appropriate software is given in the addresses section towards the end of this book.) You will not be able to experiment directly with lenses, as in a planetarium, but as these programs are capable of zooming in and out, it will be possible to measure the apparent field of view across the screen for a given magnification and choose a lens capable of covering this field.

The chances of successfully photographing selected events will be increased with simulation and with practice in handling the equipment.

Equipment

Any camera, of whatever focal length, can be used to photograph an eclipse of the Sun. Cameras with standard lenses are, however, limited to photography of ambient subjects, such as shadow bands, the oncoming shadow of the Moon, bright stars and planets around the eclipsed Sun, and the area surrounding the observer. Longer focal length lenses will be able to obtain detailed images of the solar atmosphere during the total eclipse, but exposures of a second or more will be needed. The ideal solution is to use a long-focus telephoto lens for high resolution, with wide aperture for short exposures. This may be something many of us can only dream of, given the cost of such lenses. Cameras should be fixed upon stable tripods, perhaps set into the ground. If cameras offer the option, a cable-release can be screwed into the shutter release button, which will reduce 'shutter shake' when the photograph is taken. If the lenses have a good field, the framing of totality should take account of the aesthetics of the picture. Photographs can be taken using foreground objects to add interest, with, perhaps, the oncoming shadow.

With long lenses of focal lengths between 300 and 1,000 mm, beautiful pictures of the corona can be taken. With 200–800 ISO film, and wide aperture (for example, 2.8) take exposures of ¼–2 s. If a 50- or 60-mm refractor is available, or a 100-mm reflector or long lens of focal length greater than 200 mm, interesting photographs can be secured of Baily's Beads, the chromosphere, prominences and the corona. Also, during the partial stages, these instruments can be used to image parts of the Sun, provided that care is taken to fit a solar filter or sheet of mylar in front of the objective. Eclipse photography demands easily portable, and therefore compact, instruments, and, unfortunately, their optics may have low light grasp, so long exposures will be needed.

For exposures of certain lengths, using a tripod or an altazimuth mount, and for a given focal length, the motion caused by the Earth's rotation will have to be taken into account. The critical maximum exposure time before this effect comes into play is calculated by the formula:

$$T_{exp} = 500/f$$

where T_{exp} is the exposure time in seconds, and f is the focal length of the instrument in millimetres. For example, with an unguided 200-mm lens, the maximum exposure time to avoid trailing is 500/200 = 2.5 s.

A photographic programme

With a camera attached to a telescope, or with a telephoto lens of more than 200 mm focal length, the procedure for taking a photograph during totality is as follows.

The camera, equipped with a cable release, is screwed to a stable tripod or mount. Automatic winding-on of the film is invaluable for saving time. A 36-exposure film will already be in place, and will have been tested for correct winding-on some tens of minutes before the onset of totality. Exposure (maximum) and focus (infinity) settings will already have been made, and will be checked before totality. With a film of medium sensitivity – 100–200 ISO – and focal ratio between 2.8 and 8, the exposure setting will be 1/1000 s. This will capture Baily's Beads and any prominences which appear on the limb at the onset of totality.

Several 1/1000-s shots should be taken in order to capture the moment when Baily's Beads appear. Then, a few more should be taken at 1/1000 s, this time of prominences and the chromosphere. Then, exposure settings should be increased through 1/500, 1/250, 1/125, 1/60, 1/30, 1/15, 1/8, 1/4, 1/2, to, finally, 1 s. These exposure times will capture the inner and outer corona. The outer corona will show up out to greater distances, as exposure times increase. With long focal lengths

Fig. 6.1. A large coronal mass ejection, reminiscent in shape of a tennis racquet, photographed in white light with an 800-mm telephoto lens during the 1980 Kenya eclipse. (A. Chiffaudel, ENS.)

(greater than 500 mm) the apparent motion of the Sun will be such that it will be necessary to pause to recentre it in the field of view. Finally, if totality lasts long enough, exposures should then be decreased from 1 s down to 1/1000 s.

We now examine optimum photographic techniques for each phenomenon.

First contact

For the photographer, the moment of first contact is a worrying time, because the countdown is now a visual reality. Has anything been forgotten, after all the preparations? Will the equipment be capable of capturing this beautiful spectacle?

The Moon moves across the Sun

Photographing the moment when the first bite is taken out of the Sun's disk by the encroaching Moon may be of little interest, but it can be used as a starting point for a photographic sequence. This type of photograph consists in obtaining several images of the Sun, partially and totally eclipsed, on the same photograph. To take an eclipse sequence photograph, a wide-field camera is required. The ideal focal length could be 35–50 mm, for the 36 × 24-mm format, and about 75 mm for the 6 × 6-cm format. In order to be able to reset the shutter without winding on the film, a disengagement mechanism must be present. Exposures are taken at regular intervals, between five and ten minutes, so that the images of the Sun do not overlap. Partial stages should be photographed through a filter, and totality without a filter. The difficulty in taking this type of picture lies in the framing. The Sun's path across the field of view has to be predicted (and left/right reversal of viewfinders has to be taken into account) to be certain of containing the progress of the eclipse within the frame and putting the totally eclipsed Sun in its centre. Taking the first exposure of a sequence an exact number of minutes after first contact will mean that an exposure can be taken at mid-totality, if the time is equally divided within the sequence of exposures. For example: if totality takes place at 8 h 34 m, and first contact is at 7 h 25 m, the first exposure should be taken three minutes after first contact, at 7 h 28, and the rest should be taken every six minutes, with the 11th exposure being taken at the moment of totality.

Falling light levels

A photographic record of this phenomenon can be obtained by taking spaced exposures with a lens of 28–50 mm. A film of between 50–200 ISO is best. With the same aperture and exposure settings, the same scene should be photographed as the light dims.

The advancing Moon

Photography of the partial stage is carried out at the focus of instruments, or with telephoto lenses of focal lengths between 300 and 2,000 mm. These must be equipped with solar filters which allow the use of low-sensitivity films with their fine-grained emulsions, and the shortest exposure settings to lessen the effects of turbulence.

Shadow bands

Photographing shadow bands is difficult with changing brightness levels, and can be done by following the film or camera manufacturer's advice notes. If these are unavailable, it is advisable to work at 1/125 s with a 400 ISO film at maximum aperture (smallest focal ratio).

Baily's Beads

A successful photograph of Baily's Beads is a technical achievement. Ensure that the filters have been removed to take this type of shot. An 800–2000-mm focal length is required, with emulsions of 50–200 ISO, at maximum aperture.

Second contact: totality

At this moment there is much activity, as phenomena occur in rapid succession – some of them simultaneously. Again, it is advisable not to try to photograph everything, but to confine efforts to one or two carefully pre-planned events.

The shadow arrives

A difficult moment to capture, as the shadow advances very rapidly and there are great differences in contrast between the shadow and the light.

Fig. 6.2. The corona of the 1995 eclipse, from successive images at different exposure times, combined and processed on computer at Mesai University (Japan) under the direction of Professor E. Hiei. (Mesai University.)

The darkness of totality

Superb images of the landscape and the sky, bathed in this special light, can be obtained with lenses of focal lengths between 28 and 75 mm. Exposure times will vary between 1/15 and 5 seconds, with films of 200 to 400 ISO. Simple cameras – for example, compact non-reflex types – can produce very good results. Cameras should be tripod-mounted, so that the photographer does not move the camera during the exposures.

Prominences

Photographing prominences is a fairly straightforward matter with 800–3,000 mm lenses. They are relatively bright, and red in colour, needing only short exposures. (For exposure times, according to the sensitivity of the film and *f* settings used, refer to Table 6.1.)

Photographing stars and planets

Photography of very bright stars and planets during totality can be undertaken with lenses of 28–75 mm focal length. With 400 ISO film, and with maximum aperture, results may be obtained with exposures of 1–10 s, and stars down to magnitude 9 may be recorded. Photographing less bright stars within the corona requires very good image quality, and is much more difficult.

The solar corona

Photography of the solar corona – which requires lenses of focal lengths between 400 and 1,000 mm or more to match 'professional' quality – is very problematic. This faint subject, with its extremely tenuous structures, requires high-quality images without flaws and of high resolution, and long exposures, which are incompatible with the brevity of the event.

Another difficulty arises from image turbulence due to the Earth's atmosphere, and the fluidity caused by perturbation of the wave front in front of the camera or telescope. This may be reduced when exposing by avoiding air turbulence – a result

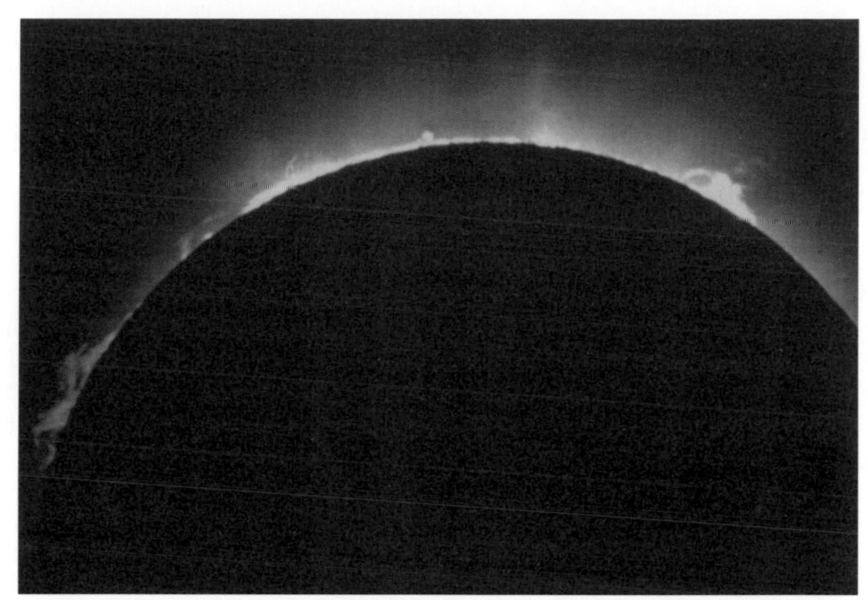

Fig. 6.3. A group of amateurs from Montana obtained this photograph of the total eclipse of 26 February 1979, using a 25-cm Schmidt–Cassegrain reflector. (E. Hattendorf Jr., and colleagues.)

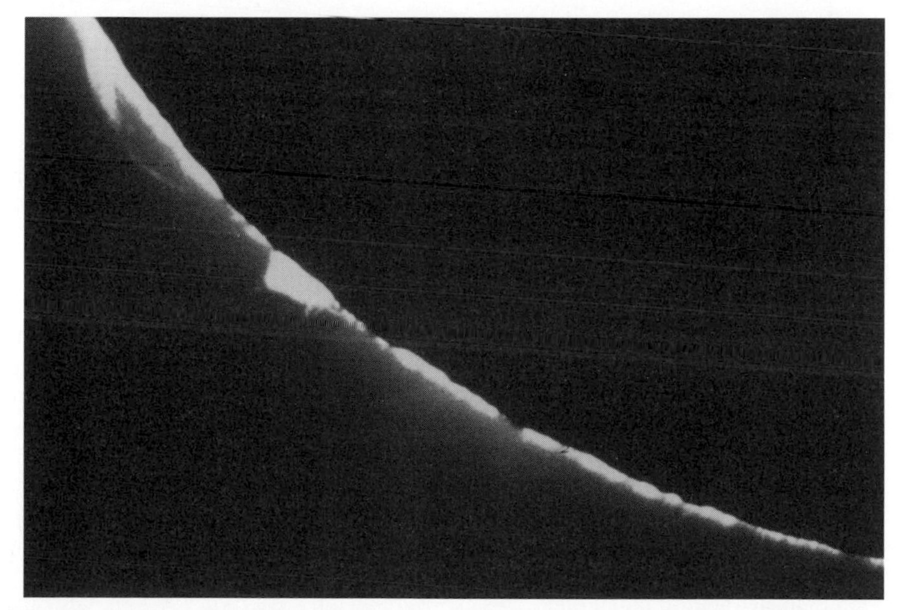

Fig. 6.4. Detail in the north-western corona during the Siberian eclipse of 22 September 1968. Original plate taken with coelostat and 10-cm objective, focal length 10 m. Note the details of the Moon's profile, and especially the chromosphere with its extensions in the form of spicules. (Kiev University.)

of cold and warm air mixing and generating motion in air masses. The effects of wind on cameras should of course be minimised, so that they can follow events accurately and secure perfect images.

Around solar minimum, the solar corona is symmetrical in relation to the Sun's polar axis. If the 36 × 24-mm format is used, it is advisable to place the north–south polar axis along the smaller dimension of the format (24 mm), since the solar plumes will be less extensive than the large equatorial streamers, which should reach out along the longer dimension (36 mm). The solar corona exhibits a strong radial brightness gradient. It is very bright near the limb of the Sun, and becomes more and more evanescent with increasing distance. It is impossible to secure an harmonious photograph showing the whole range of coronal structures. For example, if the inner corona is correctly exposed, streamers further out will not be visible. If a longer exposure is attempted, to capture the outer corona, then structures of the inner corona will be overexposed and little detail will be apparent in this zone.

To avoid overexposing the inner corona, it is possible to use a radial filter, which is designed to compensate for the corona's brightness gradient. This type of filter is made using a vacuum evaporation process, with a thin metallic layer of varying thickness being deposited on a rotating sheet of optical glass. The thickness of the deposit is calculated such that the inner corona will be attenuated, and the mean brightness gradient compensated for. As this filter is placed near the focus (in practice, just in front of the film) it must be perfectly centred on the image of the corona. It is specific to the focal length of the instrument to be used. It is, however, an expensive piece of optical equipment, and is not easy to use, so is better employed by professionals and advanced amateurs.

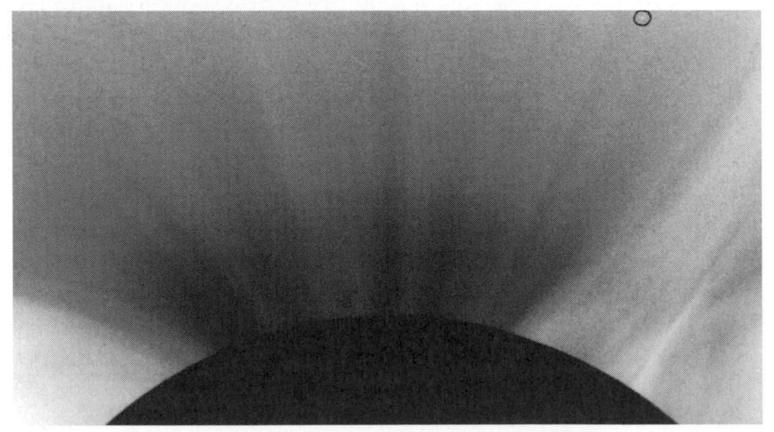

Fig. 6.5. Details from one of the coronal images taken on 30 June 1973 from Chad by the Institut d'Astrophysique de Paris/CNRS team. 20-cm objective, focal length 3 m, 15-cm diameter neutral radial filter, 20-s exposure on 24 × 18 Ekta 100 film. Only the north polar region – with a wide coronal hole centred on the pole – is shown. This region was the subject of photometric studies, with a view to constructing a model of the coronal hole, its densities and velocities. Note, at top right, a 5.7-magnitude star (circled), and, at bottom right, fine structures at the base of a large streamer seen side-on. (IAP/CNRS.)

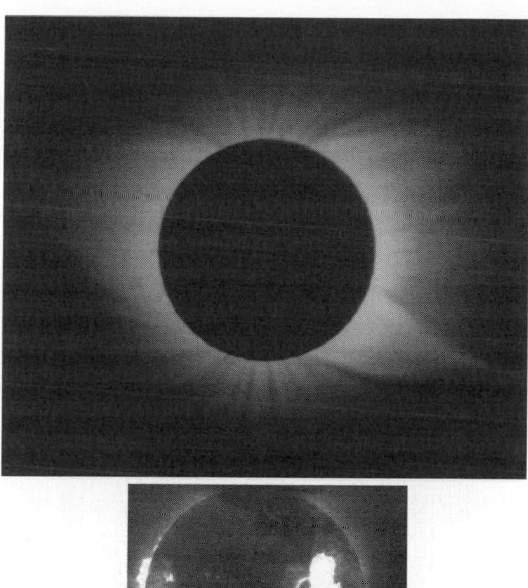

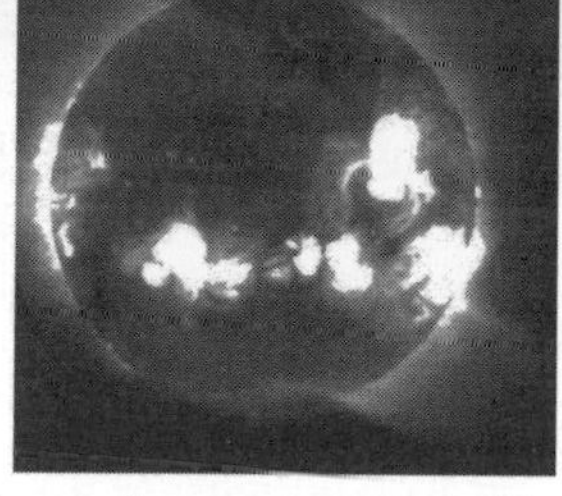

Fig. 6.6. The upper image shows the corona on 3 November 1994. Note the slight north–south asymmetry, and the large streamer to the south-west. This picture was obtained with a device using a parafocal mask to avoid overexposure of the inner parts of the corona. (F. Diego, University College London.) The lower image shows the inner corona on the same day, observed with Yohkoh's grazing incidence telescope, operating in the soft X-ray domain. Note the large coronal hole to the north, accounting for the north–south asymmetry, and the roots of the large streamer, with an apparent deviation of more than 30° from the radial direction. (Japanese Institute of Space and Astronautical Science/Lockheed Palo Alto Research Laboratory.)

But, there is a more simple technique: the inner corona can be partially masked to diminish its brightness, while allowing radiation from the outer corona to pass. To achieve this, a circular opaque mask, centred on the image of the eclipsed Sun, is placed before the focus at a typical distance of 1/3 of the focal distance, or a little less. The diameter of this circular mask must be calculated in such a way as to produce complete cancellation of the solar disk, but no more (the apparent lunar disk is a little larger). This arrangement will ensure good penumbral gradation for apertures normally used by experienced amateurs, either at f/8 or f/12. F. Diego (University College London) and A. Levy (California) achieved striking results in this way. This is also the method used by Professor Molodensky to analyse linear polarisation in the corona, and by P. Martinez in 1998. One of the authors used it in 1968, with mixed success, proving that this method is not easy. Centring the mask is the crucial factor.

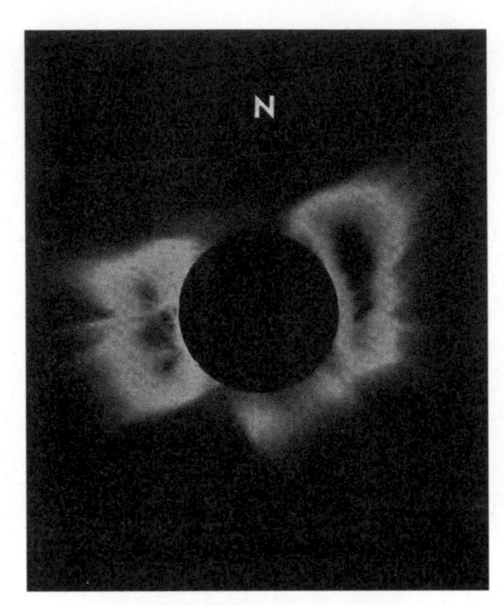

Fig. 6.7. An enhanced image of one of the photographs of the corona obtained with a neutral radial filter at Russkaya Koshka Anadyr, during the eclipse of 10 July 1972. The enhancement consisted of the superposition of a contrasted negative with a low contrast positive of the same original. This little-known process can bring out great variations in intensity between polar and equatorial regions. (Kiev University and IAP/CNRS.)

Other techniques may be used to compensate for brightness differences between the inner and outer corona. Those who process their own negatives may try unsharp masking to enhance the image. This is done by exposing, in diffuse light, a sharp negative coupled with an unused film which is slightly defocussed (unsharp), with a gap of 1–2 mm between the two films (a piece of glass serves as a spacer). The unused film will become a positive copy. The two films are then exactly superimposed, and exposed in the enlarger to obtain a paper print. This process results in the enhancement of the finest low-contrast details, while reducing background 'noise'.

Advances in computer technology have enabled amateur photographers to undertake digital processing of their images. Hours once spent in the darkroom are now replaced by clicks on the mouse and taps on the keyboard. Chemical revelations from grains of silver have become mathematical calculations depending on pixel values. It is amusing to think that the objectives of these activities remain the same: the extraction of information through elimination of blur and 'noise'.

Commercial image processing programs available include *PhotoShop* (Adobe Systems), *PhotoStyler* (Aldus) and *PaintShop Pro* (JASC). Digitisation can be carried out in a specialised laboratory, from negatives or slides, through the Kodak Photo CD process on CD-ROM. For a few tens of pounds, it is possible to store, on this 12-cm disk, up to 100 36 × 24-mm photographs in PCD format, with five resolutions – from 192 × 128 pixels at 84 Kb, to 3,072 × 2,048 pixels at 18 Mb. With flat-bed scanners, which are more and more affordable, the biggest images can be digitised. Processing programs all have the capability to enhance or attenuate the sharpness of an image. By increasing contrast between neighbouring pixels,

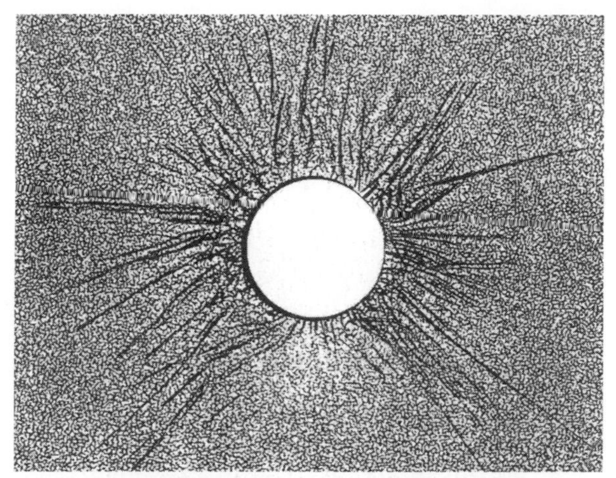

Fig. 6.8. This image of the eclipse of 22 July 1990 is a product of the *Mad Max* algorithm (maximum entropy deconvolution), an image processing program which can extract information contained in the raw image of the corona. (Kiev University and IAP/CNRS.)

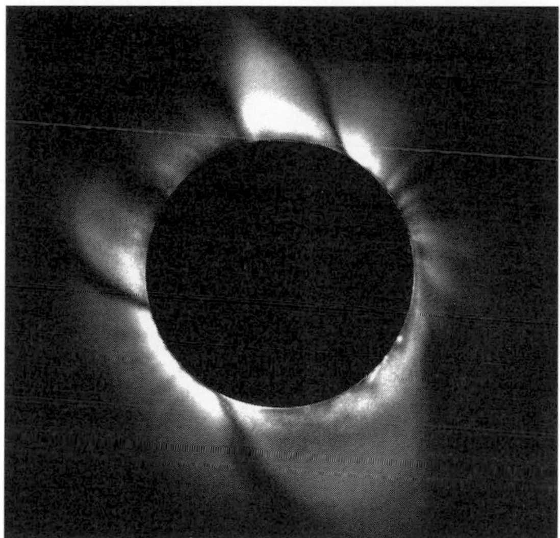

Fig. 6.9. A typical image showing what amateurs can achieve with the aid of computerised image processing. Prominences, as well as fine details in the outer corona, are visible. This montage of the 3 November 1994 eclipse was produced with a commercial program employing an unsharp mask and five images taken with different exposure times. (P. Guillermier.)

sharpness filters can improve photographed or digital image elements which are indistinct. Transitions of contrast can be softened with attenuating filters, and it is possible to create an unsharp mask from a digital image. These specialist programs have such names as *Mira* (Axiom Research), *Hidden Image* (Sehgal Corp.), *SkyPro* (Software Bisque), and *Imagine-32* (CompuScope). They are based on maximum entropy deconvolution, and have their origins in NASA research into the problems of the 'myopia' which affected the Hubble Space Telescope.

The weak link in all these computerised treatments is digitisation – the transfer from the photographic film to the digital information input medium – because this operation involves loss of information. The use of CCD technology could, however, eliminate this stage. The resolution of the latest CCD detectors available to amateurs (at the time of writing) is 1,552 × 1,032 pixels. The biggest electronic detector is 24.6 × 24.6 mm – a field of 1° 30′ × 1° 30′ for a focal length of 1,000 mm, which is worth considering when trying to include the corona. In the near future, progress will be made in the improvement of resolution, and in the field covered by the matrices; and, of course, prices will be lower. However, the problem of imaging the corona during totality remains, and, to date, CCD cameras have added little to the solution, except in the field of CCD video cameras. One of the critical points here is the 'reading time' of a CCD, which is at present prohibitively long. (To digitize for at least 14 bits/pixel, at least 15 seconds are needed, and even a few minutes for a 1 Kpixel CCD). This phase is crucial, as there is no chance of starting another image if the reader has malfunctioned, given the time it takes to verify that the image is correctly stored. Since exposure times are generally short (typically 1 s), CCDs are inefficient because of their reading times. The corona is of course a bright object, comparable to the full

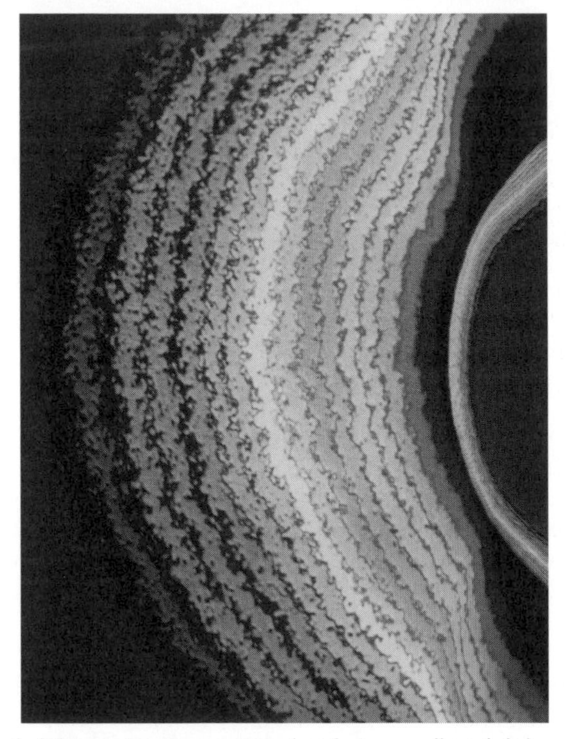

Fig. 6.10. Like height contours on a map, isophotes are lines joining pixels of the same intensity. Their forms and analysis reveal parameters linked to solar activity. (IAP/CNRS.)

Moon, and photographic films have advanced so much lately that it is difficult to achieve better results using other methods.

On the other hand, in the field of spectroscopy, and especially that involving the near infrared, CCDs have proved to be of inestimable value. This is particularly true of work involving the observation and measurement of coronal emission lines by means of a dispersal device, with or without a slit. There remains the ever present problem of 'noise' on images, which must continually be reduced without losing too much information. (Further information about CCD imaging appears in *The Art and Science of CCD Astronomy*, by David Ratledge (Springer, 1996)).

Possible comets

Photographing a field of 30° × 30° centred on the Sun, with an exposure time approaching overexposure, may place the photographer among the ranks of those who have discovered these 'hairy stars'.

Third contact

Totality ends, and, as if somebody had thrown a switch, the site is suddenly flooded with light. The eclipse photographer can now check in daylight whether the equipment has worked as planned.

The shadow departs

Like the arrival of the shadow, this is a fleeting and surprising moment, and a successful photograph of this instant is very difficult to obtain.

Reappearance of a fraction of the solar disc

Photography during this stage follows the same procedures as were described for the moments before totality. It is of utmost importance to remember to replace the filter in order to take more photographs in safety, avoiding damage both to the photographer's eyes and to the shutter diaphragm through heating.

Fourth contact

The eclipse is over. The films are rewound into their cartridges, and are labelled and stored safely in a cool, dry place before being processed as soon as possible in the laboratory.

Disappearance of the Moon's disk

This moment is of interest in that, like first contact, it can be a timing marker for eclipse sequence photographs.

Table 6.1 shows different exposure times, with allowances made for equipment used, when photographing the various stages of an eclipse of the Sun. For example, for a camera with a telephoto lens working at f/5.6, loaded with a 200 ISO film, the theoretical exposure for capturing the outer corona is 1/15 s.

Table 6.1. Photographic exposure times for solar eclipses

Film speed (ISO)	f/D								
25	1.4	2	2.8	4	5.6	8	11	16	22
50	2	2.8	4	5.6	8	11	16	22	32
100	2.8	4	5.6	8	11	16	22	32	44
200	4	5.6	8	11	16	22	32	44	64
400	5.6	8	11	16	22	32	44	64	88
800	8	11	16	22	32	44	64	88	128
1600	11	16	22	32	44	64	88	128	176
Phenomenon	**Exposure time (seconds)**								
Partial stages (1)	–	–	–	1/2000	1/1000	1/500	1/250	1/125	1/60
Baily's Beads	–	1/2000	1/1000	1/500	1/250	1/125	1/60	1/30	1/15
Shadow bands	1/30	1/15	–	–	–	–	–	–	–
Chromosphere	–	1/1000	1/500	1/250	1/125	1/60	1/30	1/15	1/8
Prominences	–	1/2000	1/1000	1/500	1/250	1/125	1/60	1/30	1/15
Inner corona	1/60	1/30	1/15	1/8	1/4	1/2	1	2	5
Outer corona	1/30	1/15	1/8	1/4	1/2	1	2	5	10
Ambient	1/2	1	2	4	8	15	30	–	–

(1) with full-aperture filter of density attenuation coefficient 4 (transmits 1/10,000 of light collected).
For the outermost corona, use the exposure times shown for ambient.
Table based on NASA publication by F. Espenak and J. Anderson.

It is easy to obtain your own souvenir photograph of an eclipse of the Sun. A photograph of use to specialist solar physicists is, however, more difficult to obtain as this type of image requires more sophisticated equipment, and a considerable knowledge of how to handle it. But, remember that specialist magazines are full of eclipse photographs, and that nothing will replace having lived through that exceptional time. Always incorporate into your plans a moment for naked-eye observation, without the constraints of technology, for a full appreciation of the spectacle.

PHOTOGRAPHING LUNAR ECLIPSES

An eclipse of the Moon can be seen from a very large area of the Earth's surface. The observer will therefore not need to travel far to observe the phenomenon. It will, however, be better to escape built-up areas, where light pollution and heat from the town can spoil observation or photography.

Techniques and equipment required to photograph, draw and film a lunar eclipse are now discussed.

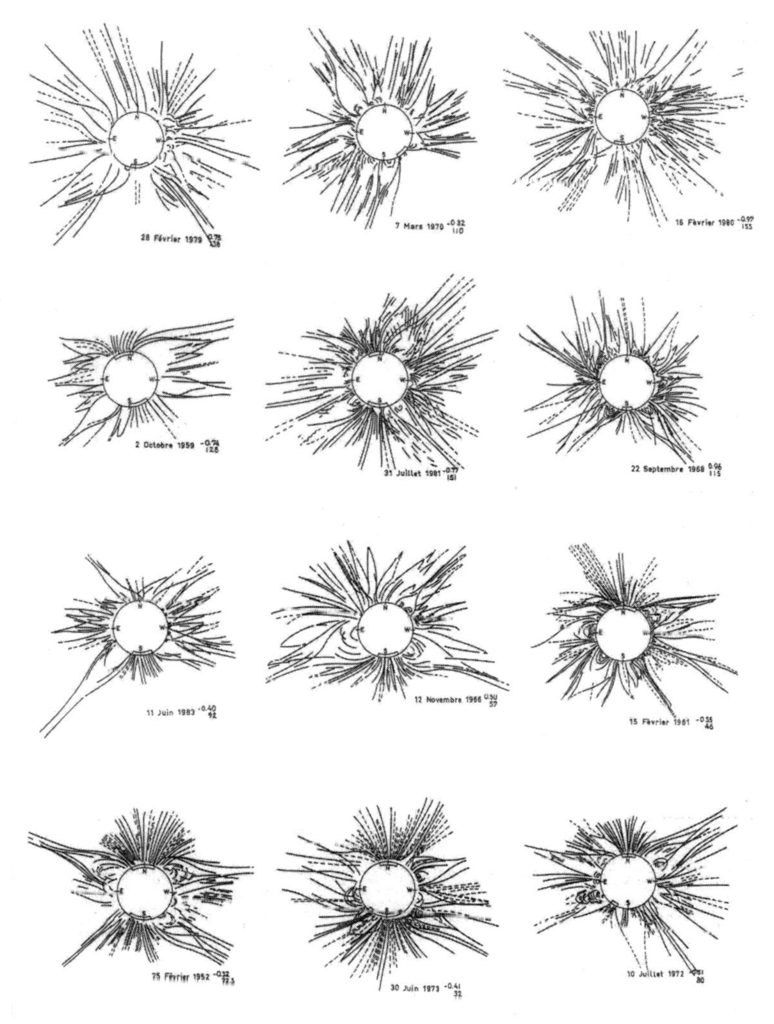

Fig. 6.11. Analysis of the global morphology of the solar corona, aided by faithful reproductions of its structures based on photographs, provides a mine of information for coronal physics, as well as insights into the cycle of solar activity. This work, carried out with the help of photo-documents using the same criteria, scales and symbols, incorporates methods and techniques attainable by amateurs. Under each drawing is the date of the eclipse, the stage in the cycle of activity, and the Zürich sunspot number. (S. Koutchmy and M. Loucif, IAP/CNRS.)

Equipment

Rule one: avoid black-and-white films. A total eclipse of the Moon is worth recording in colour. Also, as with solar eclipses, any camera, of whatever focal length, may be used. Telephoto lenses can capture detailed images of the eclipsed Moon, but their use will require exposure times of several tens of seconds.

We now discuss the capabilities of various photographic instruments, from the simple tripod-mounted camera to motorized equatorial instruments, for recording lunar eclipses.

Reflex and non-reflex cameras

Reflex and non-reflex cameras will have lenses with focal lengths of 28–105 mm, or telephoto lenses of 105–300 mm. If the shutter is left open for the whole event, a band showing the range of colour variations of a lunar eclipse will be recorded. This type of photograph can be obtained on 64 or 100 ISO film, with aperture setting 5.6 or 8, and lenses of 28–105 mm focal length. The camera should be fixed on a mount which is stable, or firmly wedged. Framing should take into account the Moon's apparent motion, making sure in advance that the Moon will be in the centre of the picture at mid-eclipse. The camera's 'B' setting should be selected, and the shutter kept open by means of a cable release.

With longer lenses of 105–300 mm focal length, it is also possible to obtain some fine images. One composition might be to frame the Moon with foreground objects, such as town lights, monuments, mountains or a forest. Full aperture should be used, together with 100–400 ISO film, and exposures between 10 s and ¼ s.

6 × 6-cm and 9 × 6-cm format cameras

With these types of instrument, several photographs of the Moon can be taken as an eclipse sequence, on the same frame. These cameras normally offer the option of resetting the shutter without winding on the film. Films used should be of 100–200 ISO, and exposure times will vary as the event proceeds. These should be planned by using the exposure times listed in Table 6.2, and adapting them to the equipment in use. As mentioned previously, the camera should be firmly mounted throughout the eclipse. Again, ensure that the Moon will be well centred on the image by estimating its motion, and timing from the beginning of the eclipse. Take one shot every five or six minutes, and the apparent motion of the sky will prevent successive images overlapping. The procession of images will show the evolution of the phases of entering and leaving the umbra, and the colouration of the Moon within the umbra. Note the time of each exposure, and the distance from the Earth to the Moon can be determined. The wide field of these formats will also show the star-field surrounding the Moon at totality.

Refractor (50–60 mm), or telephoto lens of focal length more than 300 mm

These instruments can secure images of the lit portions of the partial stages with a 400 ISO film. For totality, try 1000 or 1600 ISO, with exposure times as in the table. The formula ($T_{exp} = 500/f$) previously quoted still applies, and avoids the risk of exposing for too long and having the Earth's rotation 'trail' the image.

Refractor or reflector on a motorised equatorial mount

Telescopes may be seen as super-telephoto lenses, to which cameras can normally be attached with an adaptor often known as a T2 ring. As we have seen, the diameter of the Moon on a photographic film varies between 8.5 and 9.7 mm per metre of focal

Fig. 6.12. Total eclipse of the Moon, 3/4 April 1996. The Moon, within the Earth's umbral cone, was photographed at about 0 h 30 mUT on 4 April, through a 180-mm Takahashi Mewlon telescope, of focal length 1300 mm, using Fuji 400 ISO film; exposure, 15 s. (P. Guillermier.)

length. Detailed photographs of the Moon can therefore be taken, using 200 and 400 ISO films. However, as the exposure times are 10–60 seconds, it is necessary to drive the telescope very accurately, at the same speed as the Moon. The best method is to guide the main telescope with the aid of a guidescope mounted parallel to it and focused on a detail of the Moon.

A photographic programme

An eclipse of the Moon is not such a busy event as an eclipse of the Sun. Moreover, the time taken for the Moon to traverse the Earth's shadow can often be much longer than the time taken for the Moon to hide the Sun. Lunar eclipse photographers therefore have much more time to try different optical combinations during the various stages.

The penumbra

During the penumbral stages, photographs can be taken with telephoto lenses of 105–200 mm focal length, or at the prime focus of a refractor or a reflector. The naked eye is not sensitive enough to register the changes in the Moon's brightness as it enters the penumbra. Photographic emulsion, on the other hand, can do this. To allow comparison of the various images, though, the same exposure time and aperture should be used throughout the penumbral stage.

The Moon enters the umbra

Telephoto lenses and telescopes are best adapted to record this stage. Details of the lunar surface can be photographed as they pass from the penumbra into the umbra.

The umbra

Interesting photographs of the Moon in the umbra can be taken at all focal lengths. Wide-angle lenses (15–35 mm) can be used to create eclipse sequences. The first image will, of course, have been recorded before the Moon enters the umbra completely. Standard lenses (50–75 mm) can record the darkened, reddish Moon, and the surrounding star-field. Telephoto lenses and telescopes can be used to record details of the Moon's surface.

The above stages should be repeated in reverse order as the Moon leaves the umbra and penumbra, and the photographer will have the opportunity to obtain another set of images.

Exposure times

Predicting the colouration of the eclipsed Moon, with a view to determining exposure times, is not an easy task. Various factors have to be considered, and some of them are not definitive.

During totality, the extinction of moonlight depends upon the various factors discussed below.

Colouration of the Moon during totality

As we have seen, this is influenced by the transparency of the Earth's upper atmosphere, and especially the amount of aerosols of volcanic origin. Other factors include solar activity, and the distance from the Earth to the Moon at the time of the eclipse.

The Moon's altitude

Just as the Sun's rays redden as our local star approaches the horizon, light from the eclipsed Moon has to traverse a layer of atmosphere whose density depends upon the altitude of the Moon at the time of the event. The nearer the Moon is to the horizon, the denser will be the layer of atmosphere to be traversed, and exposure times will be correspondingly longer.

Atmospheric conditions

The air through which the light-rays pass should be very stable and dry. 'Seeing' (in the language of astronomers) refers to the degree of atmospheric turbulence. This factor can be the cause of a hazy photograph, because of turbulence, or can necessitate longer exposures if the air is misty. To avoid turbulence, it is best to observe in the open air, and not from a window or balcony. A grassy surface can be a good site, since, at night, concrete or tarmac radiate heat absorbed during the day. Instruments should have reached the ambient temperature by being placed outside at least an hour before photography begins.

Table 6.2 shows estimated exposure times for different stages of a lunar eclipse. This table is only an approximation, based upon personal experience. 'Bracketing' (utilising exposure times on each side of the value given) is recommended.

Table 6.2. Photographic exposure times for lunar eclipses

Film speed (ISO)	f/D								
25	1.4	2	2.8	4	5.6	8	11	16	22
50	2	2.8	4	5.6	8	11	16	22	32
100	2.8	4	5.6	8	11	16	22	32	44
200	4	5.6	8	11	16	22	32	44	64
400	5.6	8	11	16	22	32	44	64	88
800	8	11	16	22	32	44	64	88	128
1600	11	16	22	32	44	64	88	128	176
Phenomenon	**Exposure times (seconds)**								
Full Moon (1)	–	1/2000	1/1000	1/500	1/250	1/125	1/60	1/30	1/15
Penumbral stage before first contact and after fourth contact (2), (6)	1/1000	1/500	1/250	1/125	1/60	1/30	1/15	1/8	1/4
Near second and third contacts (3), (5)	1	2	5	10	18	–	–	–	–
Mid-totality (6)	4	8	20	40	150	300	600	–	–

Drawing lunar eclipses

For non-photographers who would like to keep a record of their observations, drawing is an alternative. Drawing the Moon is an exercise which can initiate and improve observing skills, it requires no expensive equipment, and it can reveal fine detail, unseen by photography because of the limited resolving power of films and atmospheric turbulence.

A few precautions should be taken to ensure a faithful representation.

- The observer should be comfortable, preferably seated, and the drawing paper should be fixed to a wooden board attached to the legs by some kind of elastic strap. The board should be dimly lit. The sheet may be ruled with centimetre squares to facilitate positioning.
- Drawing should commence with the sketching of larger formations, moving on to smaller and smaller details. Given the rapidity of the advancing shadow, drawing the zone nearby will mean that some of the details will either emerge from or slide into the shadow; so, an order of priority will be needed for the sequence of details to be covered.
- The lightest and darkest areas should be identified, and others, of gradations in between, can be shaded in accordingly by comparison.
- A record should be made of the date, time, instrument used, magnification and seeing during the operation.

- The drawing should be carried out without any preconceived ideas, and should not be altered after the session.

This activity provides a souvenir, as well as a precious document.

Filming lunar eclipses

It is both interesting and instructive to take a time-lapse film of the event. This can be done if the video camera has an interval meter or a time-lapse facility. An interval meter is an electronic device which operates the camera at fixed intervals. If the camera is not fitted with this accessory, it is possible to operate it manually. The camera should be on an equatorial mount, or a very stable tripod. If an equatorial mount is used, the camera should be mounted on its guiding platform. In the viewfinder, the Moon should be placed to the right of the frame at the beginning of the eclipse, about two lunar diameters from the centre. The 50- or 60-mm guidescope, of focal length 300–500 mm, with an eyepiece equipped with cross-hairs, should be aimed at a bright star near the Moon and kept upon it throughout the event. In this way, the Moon will appear to slide slowly from west to east through the Earth's shadow in the frame, as the eclipse proceeds. If, instead of a star, a bright spot on the Moon – for example, the edge of a crater – is used for guiding, then the Earth's shadow will slide across the Moon, which will be motionless in the frame. This effect is, however, less spectacular than the first. In the case of a fixed tripod, it is necessary to reset the Moon in the centre of the field after each shot.

The frequency of the shots depends upon the duration of the eclipse; the spectacle can be contained within 2–4 minutes of playback time. Filming a five-hour eclipse seven or eight frames at a time (recording for ¼ second every 30 seconds) will result in a three-minute film.

This is tedious work, needing application, and also trial filming of the Moon before the eclipse to find the most appropriate magnification. It is also necessary to become familiar with the Moon's movement in the sky. Really beautiful results can be obtained, this superb phenomenon will be better understood, and others will have the chance to see it.

Appendix A

Energy and neutrinos

What source of energy can maintain the Sun's prodigious output for such a long period? This has long been a mystery, and various hypotheses have been proposed. At first, it was thought that the Sun's energy came from chemical reactions, through burning of its constituent elements. Then, in the nineteenth century, it was believed that gravitational energy lay at the origin of the Sun's activity. Only in the early twentieth century could physicists explain the nuclear reactions which have fuelled the Sun for the last five billion years, and will maintain it for another five billion years. But certain particles issuing from these reactions still hold many mysteries.

THE ORIGIN OF SOLAR ENERGY

Let us calculate what the Sun's lifetime would be if its energy were the product of chemical combustion, rather than of gravity or nuclear reactions.

Chemical energy

Examining the chemical option, we find by rapid calculation that this is not a viable possibility. If the Sun is considered to be a stoichiometric mixture of hydrogen and oxygen, involved in the following reaction:

$$2H_2 + O_2 \rightarrow 2H_2O$$

this reaction will liberate 143 Kjoules per gram of hydrogen.

So, for the mass of the Sun and for an ideal proportion by mass of 2/18 of hydrogen, this energy is equal to

$$E_{chem} = 143 \,.\, 10^3 \times 2 \,.\, 10^{33} \times 2/18 = 3.2 \,.\, 10^{37} \text{ joules.}$$

We have seen that the Sun radiates 3.86 . 10^{26} watts; the reagents will therefore be burned away after

$$t_{chem} = 3.2 \,.\, 10^{37} / (3.86 \,.\, 10^{26} \times 3{,}600 \times 24 \times 365) = 2{,}630 \text{ years.}$$

Gravitational energy

Gravitational contraction – energy emitted in assembling all the matter in the centre of the Sun's sphere – was suggested in 1854 and 1862 by British scientist Lord Kelvin, and, in Germany, by Hermann von Helmholtz.

This gravitational energy is defined by expressing work done to assemble the mass *M* of the Sun through its volume of radius *R*, into the centre. So:

$$E_{grav} = \int G \,.\, M(r) \,.\, \delta M(r)/r$$

after integration between 0 and the solar radius,

$$E_{grav} = {}^{3}/_{5} \,.\, G \,.\, M^2/R$$

where *M* is the mass of the Sun, $2 \,.\, 10^{30}$ kg; *R* is the radius of the Sun, $6.96 \,.\, 10^{8}$ m; and G is the gravitational constant, $6.672 \,.\, 10^{-11}$ $m^3 \,.\, kg^{-1} \,.\, s^{-2}$.

So:

$$E_{grav} = 2.3 \,.\, 10^{41} \text{ joules.}$$

All the matter constituting the Sun will be gathered at its centre and will have liberated the energy E_{grav} after an interval of

$$t_{grav} = 2.3 \times 10^{41}/(3.86 \,.\, 10^{26} \times 3{,}600 \times 24 \times 365)$$
$$t_{grav} = 18.9 \text{ million years.}$$

Neither of these explanations agrees with what is observed. In fact, both the calculated lifetimes are very much shorter than estimates of the age of the solar system.

It should be pointed out that this gravitational mechanism of energy production was embraced in the late 1940s, to explain the existence and heating of the solar corona in a scenario involving interstellar micrometeorites 'falling' into the Sun. A factor favouring this model was that it explained the presence in abundance of elements such as iron in the corona, but, in the light of today's knowledge about plasma physics, it does not stand up to serious scrutiny.

Nuclear energy

What is now known about the only phenomena able to supply such quantities of energy over billions of years is largely a result of the work of the physicist Albert Einstein. In 1905, with his theory of special relativity, he demonstrated the equivalence of mass and energy, with the coefficient of proportionality between mass and energy being equal to the square of the speed of light, as expressed by his famous formula $E = mc^2$.

In the early twentieth century, Jean Perrin, who won the Nobel prize in 1926, was the first to suggest the fusion of four protons into an α particle. In his book *Les Atomes* (1913), he proposed that radioactive reactions of elements could maintain the observed luminosity of the Sun for several billion years, or even tens of billions, through the transformation of mass into energy in a reaction involving the fusion of four protons into an α particle. Only in 1938 did two American physicists, Hans A. Bethe and Charles Critchfield, suggest the ensemble of reactions which produce the Sun's energy, as well as that of other stars.

So the Sun is a gigantic thermonuclear reactor, emitting heat and light due to reactions going on at its heart, where the temperature is high enough to overcome the forces of repulsion between the nuclei of certain atoms, and maintain certain nuclear reactions.

We can again calculate the lifetime of the Sun, this time based upon the principal nuclear reaction, the exothermic transmutation of hydrogen into helium. We can write

$$4H^1 \rightarrow He^4 + 2e^- + \gamma$$

where $m_{4H} = 4 \times 1.00813 = 4.03252$ amu (atomic mass unit), and $m_{He} = 4.00389$ amu. The difference in mass is $\Delta m = m_{4H}\ m_{He} = 0.02863$ amu.

This is a relative loss of $\Delta m\ m = 0.02863/4.03252 = 0.0071$, meaning that the transformation of one kilogram of hydrogen into helium entails a mass loss of 0.007 kg. In terms of energy,

$$E = mc^2$$

$$E_{nuclear} = 0.007 \times (3\ .\ 10^8)^2 = 6.3\ .\ 10^{14} \text{ joules per kg of hydrogen.}$$

As the Sun radiates $3.86\ .\ 10^{26}$ watts, we may deduce that the mass of hydrogen converted every second is

$$3.86\ .\ 10^{26}/6.3\ .\ 10^{14} = 610\ .\ 10^6 \text{ tonnes/s}^{-1}$$

that is, 610 million tonnes of hydrogen converted into helium every second, of which four million tonnes are transformed into energy.

We may consider that, of the total mass of the Sun – which is about 70% hydrogen – only 15% of this element concentrated in the solar nucleus will be transmuted while radiation is maintained unchanged. This is a mass of

$$2\ .\ 10^{27} \times 0.7 \times 0.15 = 2\ .\ 10^{26} \text{ tonnes of hydrogen.}$$

At a rate of $610\ .\ 10^6$ tonnes per second, the reaction will last for

$$t_{nuclear} = 2\ .\ 10^{26}/(610\ .\ 10^6 \times 3600 \times 365 \times 24) = 10 \text{ billion years.}$$

It is, in fact, a chain of reactions which brings about the transformation of hydrogen into helium. It is not enough for four hydrogen nuclei to collide to produce a helium nucleus. The probability of such an encounter, even at the heart of the Sun, is very low. The transmutation of hydrogen into helium comes about through two types of reaction. The first is the proton–proton reaction (pp), responsible for 98% of the energy liberated:

$$H^1 + H^1 \rightarrow H^2 + e^+ + \nu \text{ (0.42 MeV maximum)}$$

with, sometimes

$$H^1 + e^- + H^1 \rightarrow H^2 + \nu \text{ (1.44 MeV)}$$

then

$$H^2 + H^1 \rightarrow H^3 + \gamma$$

then

$$H^3 + H^3 \rightarrow He^4 + H^1 + H^1 \text{ (85\% probability)}$$

or

$$H^3 + He^4 \rightarrow Be^7 + \gamma \text{ (15\% probability)}$$

then

$$Be^7 + e^- \rightarrow Li^7 + \nu \text{ (0.86 MeV) (90\% probability)}$$

or

$$Be^7 + e^- \rightarrow Li^{7*} \text{ (excited nucleus) } + \nu \text{ (0.38 MeV) (10\% probability)}$$

then

$$Li^7 + H^1 \rightarrow He^4 + He^4$$

or

$$H^3 + He^4 \rightarrow Be^7 + \gamma \text{ (0.02\% probability)}$$

and

$$Be^7 + H^1 \rightarrow B^8 + \gamma$$

then

$$B^8 \rightarrow Be^{8*} + e^- + \nu \text{ (14.06 MeV maximum)}$$

and finally

$$Be^{8*} \rightarrow He^4 + He^4$$

or

$$H^3 + H^1 \rightarrow H^4 + e^+ + \nu \text{ (18.77 MeV maximum) (0.00002\% probability)}$$

where ν is a neutrino, γ is a photon, e^- is an electron, and e^+ is a positron;

The second reaction is the CNO (carbon–nitrogen–oxygen) cycle, or Bethe cycle, providing 2% of solar energy.

The energy produced by the proton–proton reactions increases very little with temperature; they cannot on their own furnish the energy of massive and highly luminous stars. Hans Bethe, who had, with Charles Critchfield, already determined the proton–proton reactions in 1938, was the first to envisage other reactions, with elements heavier than hydrogen, to account for the observed energy deficit. With such elements, reactions vary markedly with temperature, as a result of their greater electrical repulsion. Bethe worked through the elements of the periodic table, and deduced a reactive cycle involving carbon which would not only maintain the reaction, but produce energy in conformity with observations. In six weeks he had finished his work and had published it in the *Physical Review* (1939). These discoveries earned him the Nobel prize for physics in 1967.

In these reactions, which also transform four hydrogen nuclei into one helium nucleus, carbon and nitrogen act as catalysts, and re-emerge intact at the end of the cycle:

$$C^{12} + H^1 \rightarrow N^{13} + \gamma$$
$$N^{13} \rightarrow C^{13} + e^+ + \nu \text{ (1.20 MeV maximum)}$$
$$C^{13} + H^1 \rightarrow N^{14} + \gamma$$
$$O^{15} \rightarrow N^{15} + e^+ + \gamma \text{ (1.74 MeV maximum)}$$
$$N^{15} + H^1 \rightarrow C^{12} + He^4$$

where ν is a neutrino, and γ is a photon.

THE NEUTRINO PROBLEM

Of the ensemble of reactions driving the Sun which have just been described, it seems that the majority of neutrinos are emitted by the initial reaction, known as the proton–proton (pp). The colossal number of protons leads to continuous generation of deuterium, making the series of reactions possible. In the latter reactions, other rarer neutrinos, and greater energy, are produced in their turn. According to astrophysical models, 65 billion neutrinos pass through each square centimetre of the Earth's surface every second. Although produced in such quantities by the Sun, these particles are extremely difficult to detect. Experiments carried out at the LEP (Large Electron–Positron Collider) in Geneva suggest that they have a mass at least 100,000 times less than that of an electron.

One of the very first neutrino detection experiments began in 1967, and was set up by Raymond Davis Jr. and his team in the disused Homestake gold mine, in South Dakota, USA. The experiment used 380,000 litres of perchlorethylene to detect neutrinos for the first time, the particles interacting with ^{37}chlorine, which was transformed by their impacts into ^{37}argon. A considerable deficit in neutrino flux was apparent, compared with amounts predicted by theoretical models. The deficit was confirmed from the outset by the Gallex experiment (so named for its gallium detector) installed in the Gran Sasso tunnel in Italy in May 1991, at a depth of 1,590 m to obviate parasitic effects of cosmic rays. The neutrino detector was calibrated by researchers from the CEA (Commissariat à l'Energie Atomique), with a 35-kg ^{51}chromium source bombarded by neutrons. The Japanese Kamiokande experiments, and the Russo-American Sage project (two kilometres below ground in the North Caucasus), confirmed the results.

All explanations of the deficit require a reworking of the theoretical model of the Sun, and the nature of the neutrino. In the first case, it is postulated that only 60% of neutrinos detected come from the proton–proton reaction, which engenders 65% of the theoretical flux. It is possible to reduce the number of neutrinos engendered by reactions other than the pp, which are very temperature-sensitive. This implies that the centre of the Sun is cooler than formerly predicted by theoretical models. Astrophysicists have suggested a number of possibilities for reducing the theoretical temperature of the Sun's core. However, helioseismology – which studies the oscillatory motions due to pressure and gravitational waves in the Sun's interior –

does not confirm the picture of a solar core where convective effects allow any significant lowering of temperature. The presence of an extremely intense magnetic field, of about 100 million gauss, could reduce gas pressure at the core of the Sun, and thereby temperature; or a greater transparency of matter than previously envisaged could facilitate the evacuation of heat produced in the core. But these are hypotheses which remain to be confirmed.

New experiments are being planned. The Super-Kamiokande, in a mine beneath a mountain in Japan, uses 50,000 tonnes of ultra-pure water, and thousands of detectors. Detection is based upon the occasional interaction of a neutrino and a water molecule. This encounter produces the emission of an electron (the Cherenkov effect). Lastly, there is the Sudbury Neutrino Observatory (SNO), operational from 1999, 2,070 m underground in a nickel mine in Ontario, Canada. Its detector is a sphere containing 1,000 tonnes of heavy water (D_2O) surrounded by 7,000 tonnes of water acting as a screen against cosmic radiation. The SNO is of interest because it is sensitive enough to detect ten neutrinos a day (50 times more than Super-Kamiokande), and can, through its use of deuterium, detect three types of neutrino. The explanation of the neutrino deficit also needs to take into account the nature of the particle itself.

Before the SNO project came into being, solar neutrinos captured by detectors were all of the same family: electron neutrinos. There are also two other forms (flavours) associated with the particles known as the muon and the tauon, and stemming from interactions with these. If a neutrino has a well-defined flavour – such as an electron neutrino emanating from the Sun – its mass is very difficult to determine. However, when this particle moves through the vacuum between the Sun and the Earth, its mass becomes more significant, for reasons connected with its dynamics; but now its flavour – the signature of its interaction with matter – is indeterminate. So the neutrino may oscillate in flight between tau-neutrino and mu-neutrino. Such an oscillation was suspected as early as 1984, as a result of experiments beneath the reactor at the EDF (Electricité de France) nuclear power plant at Bugey. Although confirmation of these oscillations has come from several experimental programmes – including those of the Los Alamos Laboratory (New Mexico), CERN (Centre Européen pour la Recherche Nucléaire) at Geneva, CNRS (Centre National de la Recherche Scientifique), and EDF, beneath the nuclear power stations at Chooz – the so-called 'neutrino problem' remains unresolved.

Appendix B

Eclipses and coronal physics

Let us try to examine the ways in which, during the last few decades, solar physicists have gleaned precious information from solar eclipses; not forgetting what we have learned from the continuous observation of the corona using space-borne instruments, building on the foundations laid by radio astronomers.

To the professional astronomer, an eclipse is a golden opportunity to secure a 'snapshot' of the Sun's 'optical' corona, with an excellent signal-to-noise ratio, due to the enormous flux of photons available. In this appendix, we describe the physical and chemical properties of this shining, shifting halo, concentrating on factors within the reach of the amateur, and on points of general interest [1, 4, 9, 15, 25]. It should be remembered that there are still many aspects which remain mysterious – among them, coronal heating and the bulk acceleration of particles. Nevertheless, for more than a century now, increasingly sophisticated methods have been employed, both on the ground and in aircraft, to analyse all parts of the corona. The absolute intensities of the white-light corona, and its emission lines and their polarisation, have been measured during eclipses, with a view to examining inhomogeneities in density of the ionised gas (plasma), and its temperature. The detailed distribution of densities may be inferred directly from analysis of fine structures in the corona, leading to an understanding of their physics through the study of the equilibrium of forces.

The investigation of dynamical phenomena, although fundamental for establishing anything like a complete description, is much more difficult during eclipses, given their brevity. However, in 1991 the 3.6-m CFH (Canada–France–Hawaii) Telescope on Mauna Kea was used to secure time sequences of small areas of the corona, and produced images of the most delicate structures yet observed. The results of this work, which may be repeated with less grand instruments, show small-scale, dynamic plasma processes very clearly. Also investigated were oscillatory phenomena and propagation waves.

This appendix therefore provides a brief account of what we know of coronal structures and the physics of the corona: its temperatures, densities, and velocities; the occurrence and structure of prominences, coronal holes, plasmoids and ejecta; and its variability, origin and evolution. Also included are a few simple considerations of what remains to be done to add to our knowledge of the physics of this extraordinary 'nebula', which we see periodically during total eclipses.

AN OUTLINE OF THE PHYSICAL STUDY OF THE SUN'S CORONA

We are not attempting a study of the solar corona here, as this would be difficult to fit even within a whole book. The subject has been amply covered in a recent work, aimed at university students, by two outstanding specialists [1]. It includes a list of more than 700 articles on coronal physics.

We shall therefore limit ourselves to an overview, mainly for confirmed eclipse observers, but also (and why not?) for all those who like to contemplate the whys and wherefores, and even the meaning, of this ever-changing cloud.

During a total eclipse of the Sun, when first the photosphere and then the chromosphere are completely occulted, there appears a partly polarised white halo, produced essentially by the scattering of the light from the Sun's disk by free electrons (K corona). This is the white-light plasma corona, responsible for mass loss via the quasi-radial flow of gas (solar wind), the origin of which is still not completely understood. Thus, surrounding the moving disk of the Moon, and slightly larger than the eclipsed Sun, the corona is visible, from the transition region to the limits of the most elongated structures – for example, streamers and plumes. The great streamers, burgeoning out to distances of millions of kilometres, are the most striking structures. They tend to be confined within heliospheric 'sheets' (Figure B.1), especially at solar minimum.

With medium spatial resolution, images and spectra may be obtained in order to analyse of densities, temperatures and velocities. During a natural eclipse, there are no limits to the field of view, and, in good observing conditions, any limitation on the accuracy of measurements is due to the well-known and constant brightness of the dusty F corona [2, 10]. This dust component must obviously be subtracted, with the aid of the model shown in Figure B.7, so that the K (plasma) corona can be studied. The more or less constant brightness of the eclipse sky can be reduced by observing from an aircraft, or by differential methods. This brightness varies from one eclipse to the next, according to the magnitude of the eclipse, weather conditions and the altitude of the Sun. The sky background is also polarised, if only slightly.

During eclipses, absolute photometric measurements of the corona can be carried out by using images of sufficiently bright reference stars (equal to, or brighter than, visual magnitude 8.5), studied simultaneously on the same image as that of the corona. The Sun's light, suitably attenuated, may also be used for standardisation purposes, if no stars are visible on the images. The aim is to determine the intensity at each point in the corona, in the same frame of intensity as that of the 'mean' disk.

In recent decades, progress has been made partly through the use of the neutral-density radial-gradient filter, the accurate gradations of which compensate for the considerable mean radial variation [2] of the white-light corona, and by eliminating parasitic effects from the instrument and from the sky (the halo effect). The use of such filters appreciably improves resolution, avoiding overexposure and reducing dynamics.

Fig. B.1. Wide-field images obtained by the Los Alamos group during the eclipses of 16 February 1980 and 10 July 1972. North is at the top, and east to the left. Note that the masking disk superimposed by computer overlaps the Sun. On the image of the 1980 eclipse, the relative size of the Sun is indicated. (Montage by IAP/CNRS, from original images processed and kindly lent by C. Keller, Los Alamos Laboratory.)

For the 1991 eclipse, highly linear and ultrasensitive detection systems were used: namely, digital CCD cameras, especially for time sequences with narrow-band filters. However, photographic film is still widely used for eclipse work, due to its simplicity, reliability and information storage capacity.

In the infrared, 2D matrices were used for the first time in 1991, in order to study the outer corona (beyond a radial distance of 2.5 solar radii from the centre of the Sun). In this case, the radial filter was replaced by an external ('double') occultation system [26].

Optical and near-infrared spectroscopy can also be attempted during total solar eclipses, and numerous coronal emission lines are easily seen with, for example, a slitless spectrograph (Figure B.2).

Nowadays, much can also be gained from the use of modern instruments for more extensive investigations which cannot be undertaken by other means. For example:

- Analysis of the profile of lines from less bright areas, where it is thought that the solar wind is more rapid; for example, in coronal holes, and also in streamers at distances of more than two solar radii.
- The study of Fraunhofer lines diffused by interplanetary dust and especially by the electron component. Because of the lack of photons, X-ray and EUV

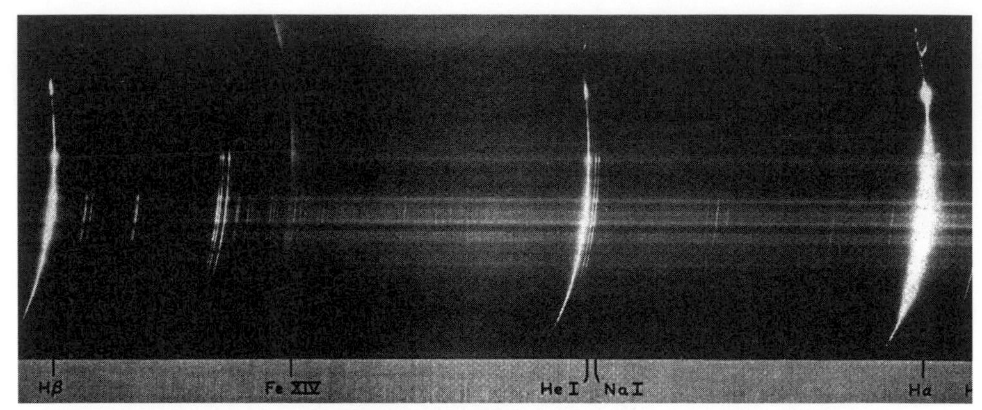

Fig. B.2. This 'slitless' spectrum of the innermost corona and the transition region between the chromosphere and the corona was obtained by a Japanese team from Kwasan Observatory two seconds after the beginning of totality during the eclipse of 7 March 1970 in Mexico, at 17 h 27 m 18 s UT. The system comprised a grating of 600 lines per millimetre, and the spectrum covered wavelengths of 390–810 nm, with a mean dispersion of 2 nm mm^{-1}. Emission wavelengths identified are Hβ at 486.1 nm, FeXIV at 530.3 nm, HeI at 587.6 nm, NaI at 589.0 and 589.5 nm, Hα at 656.3 nm, and HeI at 667.8 nm. (Kwasan Observatory.)

(extreme ultraviolet) spectroscopy from space can hardly obtain precious accurate information, even from analysis of a true emission line at distances greater than two solar radii. Analysis of such space observations is possible only after long hours of integration, which dampen effects and do not favour the measurement of 'instantaneous' velocities.

- The Doppler effect: firstly, study of the profile, much widened by velocities, of strong Fraunhofer lines favours the direct measurement of the electron temperature; secondly, analysis of (forbidden) emission lines allows direct measurement of the velocity of plasma motions, using the Doppler effect. The width of the lines (for measurements of the velocities of disturbances), and even the radial gradients, are not well known beyond a distance of 1.5 solar radii, where waves are prominent. Because of parasitic scattered light, even the Lyot coronagraph is not well suited for this work.

Total eclipses therefore provide an opportunity to verify, through advanced investigation techniques, theory on heating and acceleration of coronal plasma (the solar wind).

Unfortunately, total eclipses are rare and short-lived events. Also, coronal activity differs from one eclipse to another, and it is difficult to verify the results of a 'snapshot'. For this reason, results from the most sophisticated experiments are sometimes disappointing. Finally, no information at all about the solar disk is gained during an eclipse. Fortunately, instruments working in the soft X-ray domain, and EUV/optical coronagraphs in space, provide very extensive observations which complement what is derived from eclipses.

The results of white-light polarimetry and photometry, and also of structural analyses from eclipses, are nowadays well established, and cover many solar cycles [3]. They have made an indispensable contribution to the analysis of the cycle of coronal activity. Occasionally, even coronal mass ejections (CMEs – an ingredient of coronal activity) have been observed during eclipses. They are created from 'magnetic explosions' in the inner corona, within large-scale structures which become, more or less suddenly, unstable. For obvious reasons, space-borne coronagraphs, with their more extended coverage in time, are far more suited to observe CMEs from orbit. But the best observations of the corona, from the point of view of spatial resolution, were made with the 3.6-m CFH (Canada–France–Hawaii) Telescope on 11 July 1991, and they cast new light upon the study of very fine phenomena – a study which has great bearing on the physics of small-scale so-called 'turbulent' processes [27], where energy of magnetic origin is dissipated. So we can certainly see CME-like phenomena during each total eclipse, but on a small scale of a few arcsec.

MORPHOLOGICAL ANALYSIS OF CORONAL STRUCTURES

Theory

Detailed analysis of the quasi-stationary structures of the corona (which includes to a certain degree, fine structures) advances our knowledge of solar magnetic fields, and hence of the Sun's activity and cycle.

Solar physicists define a parameter β, which characterises solar plasma. β is defined as the ratio between the density of kinetic energy of the gas (or pressure *P*) proportional to the product N . T (N being the density of particles of mean mass *m*, and T kinetic temperature), and the energy density of the magnetic field, proportional to the square of the amplitude of the 'local' magnetic field.

So

$$\beta = 8\pi P/B^2 \tag{B.1}$$

where β is the ratio of energy densities; *P* is the local gas pressure of plasma; and *B* is the magnitude of the local magnetic field.

A plasma of low β (with magnetic 'pressure' dominating gas pressure) would 'align' itself along magnetic field lines, even if the corona were in its dynamic scenario. It can be shown too that β is also the ratio between the squares of the two very characteristic velocities of plasma: the speed of sound c_o and the Alfvén velocity v_A at which magnetic disturbances move ($\beta = c_o^2/v_A^2$). Flows of gas occur along the magnetic field lines, as the magnetic force always acts at right angles to the field lines. In reality, the situation is more complex: for example, there is a force of tension of the field lines when they are curved or twisted.

Turbulence is also a factor, especially on the small scale. The high levels of electrical (and thermal) conductivity of the coronal plasma are, in the final analysis, a microscopic determining factor, tending to 'hold' the gas on the field lines, which may nevertheless be torn away by 'exterior' activity such as explosions and reconnections.

On eclipse photographs, structures upon the regions of coronal holes appear open, as do the ambient magnetic field lines. Conversely, arched and looped structures are clearly seen around active regions and filaments [4] corresponding to closed magnetic field lines which are completely disturbed or even detached during a CME event.

To complete this brief description of the equilibrium of forces in the coronal plasma, we must include the omnipresent solar gravity g directed radially towards the Sun's centre, and the decreasing radial gradient of gas pressure ∇P, acting in the opposite direction.

Ignoring the magnetic force, we may express this as $\rho g = \nabla P$, and define a scale of height h in the atmosphere, with $h = \mathrm{k} \,.\, T/M \,.\, g$ (k being the Boltzmann constant). To calculate the mean density ρ of particles of mass M_i in the ionised gas, note that account must be taken of the densities N_i of all the elements, electrons and ions ($\rho = \Sigma_i \, M_i \,.\, N_i$). Much of coronal physics is concerned with magnetohydrodynamics (MHD), to which a good introduction will be found in [1], and in [9] by J. Heyvaerts and E. Priest.

Practical aspects

Photographs of eclipses taken in white light do not show the numerous details of the corona correctly, unless special filters are used. The most sophisticated method involves the use of a neutral-density radial-gradient filter, because of the strong radial gradient of the light, especially in the intermediate corona. In this case, details

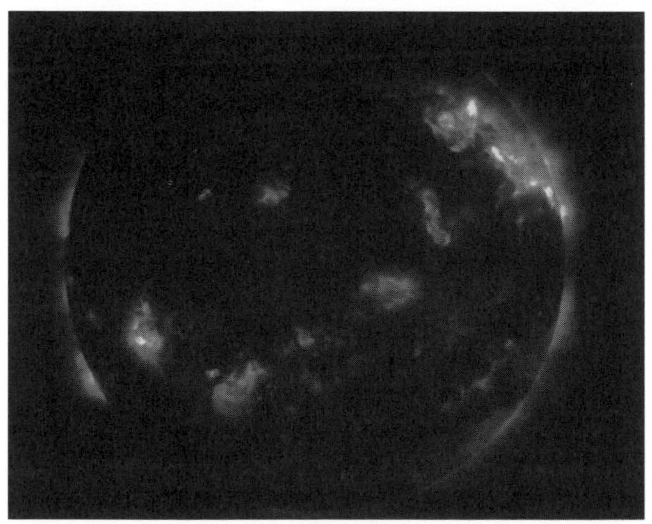

Fig. B.3. X-ray image of the solar corona obtained on 11 July 1991 by a rocket launched from New Mexico as the eclipse reached totality in Hawaii, during experiments by L. Golub's group from the Center for Astrophysics/Smithsonian Institute and IBM-USA. Note the effect of the edge of the Moon, at far right, projected against the corona behind it. Resolution on the original images is excellent, and unfortunately cannot be reproduced on this scale. (Smithsonian Astrophysical Observatory and Center for Astrophysics, Harvard, USA.)

of the corona appear more clearly, as long as a sufficiently long focal length of a few metres is used. However, because of the superimposition of details along the line of sight, as well as the considerable differences in brightness between the polar and equatorial zones, photographs will not reflect what an experienced observer might see. The image reproduced in [2], [3] and [8] of the eclipse of 30 June 1973 shows that an image reprocessed on a computer may approach in quality an interpretation drawn by a naked-eye observer. A white-light image normally amply reveals the distribution of densities in the corona, since the amount of light observed is directly related to the quantity of free electrons; and, as a consequence, with plasma densities integrated along the line of sight, due to the existence of electrical forces in a plasma, on the average the density of electrons is equal to that of ions. An examination of this type of image shows that the plasma is essentially confined to fine structures, such as filaments and loops. This impression is confirmed by the study of an X-ray image obtained from space, showing emissions in the inner corona observed against the solar disk (Figure B.3). These emissions are very sensitive to high temperatures.

Plasma densities can be deduced directly from white-light observations (Figure B.4). Thermal structure is studied from filtergrams taken in the wavelengths of 'forbidden' emission lines. The richness and complexity of structures is readily apparent. This suggests that dynamic phenomena are essentially small-scale occurrences, while abrupt instability states operate on the large scale. Although these numerous observations embrace all orders of size, better resolution will be necessary for a detailed understanding of the physics of these structures, and an answer to the question of the origin of heating.

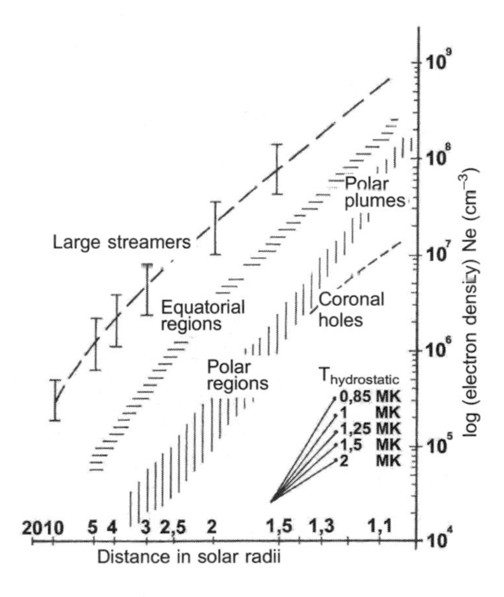

Fig. B.4. Radial variations in plasma densities in the corona, for typical structures or regions. Scales are set to indicate according to the slope of hydrostatic temperatures. (IAP/CNRS.)

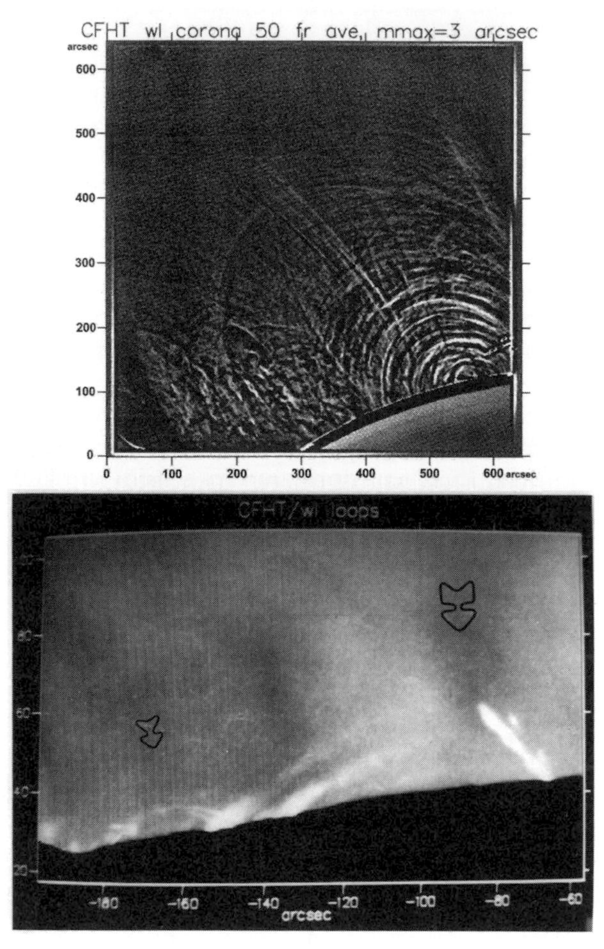

Fig. B.5. Details from images of the solar corona obtained with the 3.6-m Canada–France–Hawaii Telescope on Mauna Kea, Hawaii, during the total solar eclipse of 1991. The upper image shows an inextricable tangle of loops brought out by the algorithm *Mad Max*, specially developed for treating coronal images. On the lower, untreated, image, note the Moon's edge, and especially the very bright structures (arrowed), which are tiny prominences, very difficult to observe when the Sun is not totally eclipsed. (IAP/CNRS and SPO/NSO.)

Numerical models can also help in the analysis of the morphology of the corona.

Resolution was improved [29] by processing a 70-mm film taken in white light with the 3.6-m CFH Telescope, achieving the best resolution ever for images of the corona (Figure B.5). However, a digital CCD video camera, working with a very small field, managed even better resolution at the focus of the same telescope, but on another region of the inner corona; coronal 'fibrils' only 0.4″ in diameter were measured [27], confirming the theoretical predictions of H. Alfvén [19].

Discussion

If we consider the equilibrium of pressure in coronal plasma in the radial direction, and assume an isothermal corona, pressure P of ionised gas in structures can be taken to be approximately equal to the pressure of the magnetic field. This equality is at least true for the general corona, where P varies as r^{-6}. The amplitude B of the large-scale magnetic field (first harmonic) of the photosphere decreases like that of a point dipole, as a function of r^{-3}, in this range.

With the help of the 'potential field' approximation, and with appropriate limits ascribed to conditions, numerical methods have been applied in order to extrapolate values for the magnetic field at the level of the photosphere. This approximation involves considering only those fields induced in the corona by surface magnetic monopoles, as if the corona were devoid of plasma and therefore of currents. The resultant global morphology is fairly comparable to the observed density structures, if details are ignored, and if field lines are forced to 'open out' at great distances.

On the very large scale, this approximation is correct [5, 6]. But the corona abounds in very fine structures like jets and filaments, and the β of the plasma, being variable, is not known. Moreover, only conjectures can be made about the magnetic field; such is the case, for example, with coronal discontinuities occurring further out [28].

Accordingly, a fairly detailed knowledge of the distribution of plasma densities is a prerequisite for calculating the magnetic field which could counterbalance the lateral gas pressure. Loops are a special case, as account must be taken of dynamic equilibrium, and of the fact that the obviously 'twisted' field is also non-potential. Electric currents must, therefore, also be considered. It will be necessary to integrate these currents and the resulting magnetic forces. Inhomogeneities of temperature play a hidden rôle here: in loops, there prevails a kind of law of proportionality linking temperature, pressure and the length of the loop.

At first sight, it is tempting to consider that temperature is responsible for the variation in the height of arches seen in the mid-corona, outside or in proximity to active regions. Appreciable variations are noted during the solar cycle. Radial variation in mean density during the cycle is no doubt linked to that of temperature through the scale of height. However, arches are visibly bigger at the time of sunspot minimum, when the temperature of the corona – with its supposed links to the intensity of the solar magnetic field – is imagined to be lower. This runs counter to the law of scale in the loops. We may therefore suppose that the morphology of the corona doubtless depends more upon the evolution of the Sun's large-scale magnetic field, and has no direct relationship with thermodynamic parameters such as temperature, or even perhaps the quasi-static magnetic field, but with their temporal variations on quite small scales. This may be expected, for any mention of magnetic field variations implies induction, and in turn an electrical field and current.

The existence of numerous loops above and in proximity to active regions strongly suggests that the coronal plasma is 'frozen in'. This means that it is trapped within field lines, if β is small with, as a consequence, very high electrical conductivity (supposedly infinite), and therefore an absence of resistivity. Here, the plasma is considered to be a continuous fluid. This approximation, which, let it be said, is very useful, is often poorly applied and leads to mixed results – as H. Alfvén, who introduced it, pointed out. As with prominences, there are many places in the mid- and outer corona where densities are very accentuated above neutral lines of magnetic polarity reversal, which extend outwards from active regions. Extreme cases are the great equatorial streamers at times of sunspot minimum. It is thought that this equatorial streamer 'belt' is the source of the phenomenon known as the heliospheric sheet [6], with its remarkable folds [30]. It should be noted, though, that streamers also appear at higher latitudes, and even at the poles, at times when the dominant polarity of the magnetic field is reversing. It is appropriate, therefore, to take into account not just the solar cycle, but also the morphology of the corona [3]. In [3], [9] and [24], the deviation of the main structure of streamers from the radial direction is explicitly a function of the phase of the cycle. The same result is obtained by taking the Wolf Number as a parameter [3].

By assuming that coronal plasma is confined within streamers and sheets above neutral lines [7, 8, 30] of the chromosphere, we can now produce a simplified representation of the corona which agrees quite well with the actual distribution of densities in the outer corona. This representation differs appreciably from the map which may be deduced from calculation of the 'magnetic field lines' based on the 'current-free' approximation (potential field) and on the observation of magnetic fields at the level of the photosphere, as was shown by the Russian theoretician M. Molodensky [30].

Detailed analysis of these maps [5] shows that they reflect only very approximately the individual structures actually observed in the outer corona. The existence of very steep azimuthal or tangential pressure gradients – much more intense than radial gradients – also hinders interpretation in terms of the potential field.

To explain the steep transverse gradient of local density observed in the corona, we know of only one force capable of counterbalancing the corresponding pressure forces: the magnetic force. This latter equals $J \wedge B$ (where $\wedge$ is the vectorial product, J the electrical current density and B the magnetic field). This force is sometimes known as the Ampère force (when electric currents are involved), the Laplace force, or the Lorentz force (when only isolated charges are being considered). As J and B are vectors (functions in three directions) this force is quite difficult to calculate normally; but, in conformity with Ampère's law, it is strictly perpendicular to the field lines, which means that not too much work is involved.

It is not easy to measure electric currents within a plasma. To attempt to 'see' what is happening, it is preferable first to consider the magnetic field induced in the corona by the most powerful currents, which are those due to motions in the dense layers of the photosphere beneath.

To obtain an approximation of the field induced above an active region simply composed of sunspots and faculae, we use a classical axisymmetric model with a double ring of coplanar currents, which facilitates precise calculation of the magnetic field outside it. The structure arising from this calculation is quite similar to typical structures observed in the corona (Figure B.6).

A point of interest here is the existence of special separatrix surfaces, where the

Fig. B.6. Magnetic field line structures calculated using a simple axisymmetric model of rings of electric currents in opposite directions, to simulate a sunspot and its neighbouring faculae. The magnitude of the field appears in shades of grey, and the region around the singular point where the field cancels out (a magnetic trap?) is darkest. Note the similarity of the shape in this simulation to structures observed in the corona during eclipses. (Numerical model by O. Koutchmy, Paris VI, and IAP/CNRS.)

gas is concentrated, and a singular or neutral point, where the force of the magnetic field disappears, as the field is nullified there. This region can therefore be considered as a zone of confinement for coronal plasma. In certain circumstances, a quite abrupt annihilation of magnetic energy can occur within this zone; for example, if rapid configurational modifications are caused at the base of the field lines. But here we are entering an original domain of plasma physics, which has a bearing on laboratory research into controlled nuclear fusion, and which is largely outside the scope of the present account. It should also be remembered that, in the case of cosmic plasmas, gravity and its radial gradient are omnipresent.

QUANTITATIVE PHOTOMETRIC ANALYSIS OF DENSITIES [2, 11, 13]

General case and homogeneous case

The absolute intensity (expressed in units of the well-known radiance of the solar disk at the same wavelength) of the white-light corona is linearly linked to the densities of free electrons, which is also the density of the plasma, as the medium is, on average, electrically neutral and electron density is equal to that of the ions. The scattering (Thomson scattering) of 'optical' photons by free electrons in the plasma is relatively isotropic, but differs by a factor of two when following orthogonal directions, and therefore planes of polarisation of the radiation.

If θ is the angle of scattering, the coefficient of Thomson scattering may be written thus:

$$\sigma = \sigma_0 (1 + \cos^2\theta) \tag{B.2}$$

where σ_0 may be taken to equal $0.66 \cdot 10^{-24}$ cm^2, or about ten times the effective section of an electron. Eclipse photometry can therefore enable direct measurements of coronal densities, although accurate calibration of the intensities is still necessary.

The contribution of the F corona – produced by scattering from solid dust particles, much further out but in the line of sight – is known (Figure B.7). This dust is of silicate type, with a mean particle diameter of 10 μm. The particles orbit the Sun, and their density seems to be greater, the nearer they are to the Sun. On an image of the solar corona, these particles – seen by grazing light from the solar disk –

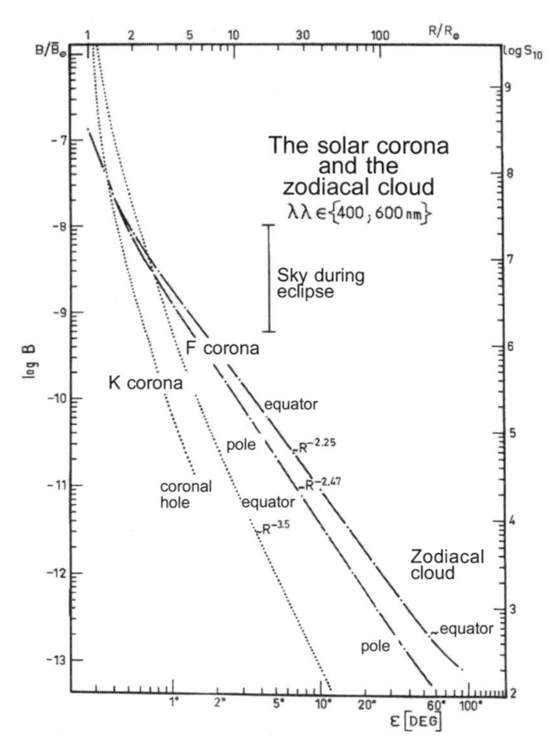

Fig. B.7. Radial variations in intensities observed in the eclipse corona, in the visible optical domain. K is the high-temperature plasma component composed of structures; F is the dust component, more or less invariable over time, which merges into the zodiacal component seen at night. Units of intensity are: 1) at left, $B/B_\odot$, in mean brightness of the solar disk at the same wavelength; or 2) at right, S_{10}, which is the count of stars of magnitude 10 per square degree of sky. Scale units are expressed thus: at the top are radial distances in units of solar radii, and at the bottom, in angular distance from the Sun's centre, in angular degrees. (S. Koutchmy and P. Lamy, IAP and LAS/CNRS.)

produce an homogeneous halo (F corona) which must be carefully subtracted to distinguish the plasma (K) corona. It can be shown that the low-angle scattering from the dust particles is principally a product of diffraction and reflection, and is somewhat similar to Thomson scattering, which is more isotropic (see equation B.2).

The atmosphere of the corona is roughly homogeneous, and measurement of the integrated degree of polarisation can separate out the K and F components; the F corona is not polarised at distances less than five solar radii [10].

The 'A' background, due to parasitic light from the instrument and the sky, is also not considered to be polarised, and the degree of polarization P_K of an homogeneous model of the K corona can easily be determined [11] by means of the theory of Thomson scattering.

The observed linear polarisation deduced from measurement of radial brightness B_R and the tangential brightness B_T of the corona is equal to

$$P_{Tot} = (B_T - B_R)/(B_T + B_R)$$
$$P_{Tot} = (K_T - K_R)/(K_T + K_R + F + A) \quad (B.3)$$

and

$$P_K = (K_T - K_R)/(K_T + K_R) \quad (B.4)$$

The brightness of the K corona alone is thus deduced as

$$K = (K_T - K_R)/P_K$$
$$K = P_{Tot}/P_K \,.\, (K + F + A) \quad (B.5)$$

from measurement $K + F + A = B(r)$, from P_{Tot} and from theoretical calculation of P_K. Note that mention is made here only of linear polarisation characterising the amplitude of luminous vibrations following two orthogonal directions, and the preferred direction of these vibrations. For Thomson scattering, this direction is strictly tangential (electrical vector) to the solar limb beneath.

This approach was largely developed in the past to describe observations and to deduce mean density models [2, 11, 13, 14]. As long as the data are well calibrated, it is possible to deduce the contribution of the F corona using the data available in [10], as this component is constant, and the accuracy of model [10] is better than 10%. This obviates the need for polarimetric analysis, which is tedious, though extremely interesting in the analysis of the structures of the plasma corona.

Although mean values for distribution of electron density are useful for comparing the corona at different latitudes, or at different times of the sunspot cycle, they are far removed from real values. The most conservative estimate of the filling factor in the mid-corona gives a value of less than 10%. Mean values therefore produce a simplified picture of the corona, with no mention of really fine structures of diameters less than 10 arcsec. The study of the actual distribution of density in the structures necessitates complex formulae comparable to those employed in tomography (for example, inversion of the Abel integral with various kernels). Here, we follow the classic approach of the homogeneous case, after Baumbach [11].

Consider the surface brightness of the K corona (intensity as measured on images) – $I = I(\rho)$ – along a radial direction where ρ is the projection on the plane of the sky of the real radial distances. It is convenient to represent I(ρ) by a sum of terms in the form ρ^{-m} (*m* having different values between 17 and 3.5 according to the value of the radial distance). Between the surface and a distance less than three or four solar radii, it will be seen that the inferred radial densities of the ionised gas, for $m_1 = 17$ and $m_2 = 7$, follow a curve described by a hydrostatic law with a fairly constant temperature T_H of $1\text{–}2 \times 10^6$ K (10^6 K = 1 MK). This seems to indicate that the departure from hydrostatic equilibrium caused by the hydrodynamic flow (solar wind) is small in this region.

Let us try to translate this by means of equations. Supposing that: μ – the mean mass of coronal particles (electrons, ions such as protons and α particles, and more massive nuclei), in units of proton mass M_H, 1.66×10^{-24} g – is known, after consideration of the abundance and degree of ionisation (H and He are entirely ionised and, at first approximation, μ = 0.6); G is the gravitational constant at the solar surface; then

$$N_e(r) = N_e(R_\odot) \,.\, \exp[-(\mu \,.\, M_H \,.\, G \,.\, R_\odot/k \,.\, T_H)(1 - 1/r)] \qquad \text{(B.6)}$$

where k is the Boltzmann constant, equal to 1.38×10^{-6} erg . deg^{-1}; and $N_e(R_\odot)$ is the coronal density of electrons at the Sun's 'surface'.

A more convenient formula may be deduced [16]:

$$N_e(r) = N_e(R_\odot) \,.\, \exp - (13.9/T_H \,.\, h/r) \qquad \text{(B.7)}$$

where T_H is the hydrostatic temperature expressed in MK, constant at first approximation; *r* is the radial distance in solar radii; and *h* is the altitude above the photospheric surface expressed in the same units.

Seemingly, this practical formula is valid only for the inner corona (with no hydrodynamic flow), assuming T_H is known, as T_H can be found from knowing the radial gradient of $N_e(r)$ from images of the corona.

If we suppose that the coefficient of filling of the corona is constant, or, of course, if we postulate that the corona is homogeneous and stratified, then $N_e(r)$ is calculated from measurement of I(ρ), which is the mean brightness of the corona at radial distance ρ projected on the plane of the limb, by inverting the following formula:

$$I(\rho) \approx B_\odot \sigma_0 \,.\, \int N_e(r) \,.\, W_\lambda(r) \,.\, (1 + \cos^2\theta) \,.\, dl \qquad \text{(B.8)}$$

where σ_0 the isotropic part of the coefficient of Thomson scattering (see B.2); $B_\odot$ is the mean brightness of the solar disk; $W_\lambda(r)$ is the dilution factor taking account of limb darkening on the solar disc; θ is the angle of scattering used to take account of anisotropy of scattering; *l* is the length of line of sight.

With a good approximation we may suppose that $W_\lambda \approx W_\odot/r^2$ and $\langle (1 + \cos^2\theta) \rangle \approx 1.25$, because the effective length of integration of the line of sight l_{eff} is defined as the length of the line of sight corresponding to a decrease in intensity of the quantity of light reduced by a factor e = 2.7, compared to the value obtained in the plane of the sky. Nevertheless, inversion of (B.8) is questionable, as structures are involved. Taking a radial variation of $N_e(r)$ given by equation (B.6), and choosing a hydrostatic temperature T_H of 1.5×10^6 K, l_{eff} can also be determined. Accurate values are given in [29].

The case of large structures

We can now introduce the filling factor in its simplest form. Supposing that along l_{eff} only a part α of the line of sight is occupied by plasma at constant density, then the filling factor is equal to α. When the intensity of coronal emission lines or radio emission are considered, another factor – often brought in to describe the irregularity of the density – is $<N_e^2>/<N_e>^2$. There is a relationship between this factor and the filling factor. However, since there exists a whole range of irregularities in the corona, and observations have finite resolution, this relationship is not easy to determine or measure.

Note that the allowed lines of the extreme ultraviolet, for example, as with radio emissions, are also a function of local coronal temperature, and not of N_e^2 only. The emission of forbidden lines in the visible is also a function of a power of nearly two of the electron density. (To ensure the electrical neutrality of the plasma, the density of the electrons is everywhere the same as that of the ions.) As previously stated, analysis of the degree of polarisation P_K in the K corona leads to a better understanding of the distribution of N_e along each line of sight. Let us examine this point. The greatest polarisation is produced by electrons at an orthogonal angle of reflection ($\theta = 90$). Outside this plane (limb plane, or plane of the sky, defined as the plane at right angles to the line of sight and passing through the centre of the Sun), polarisation decreases. Assuming an homogeneous axisymmetric distribution, and T_H to be 1.5 MK, the degree of polarisation of the K corona can theoretically be calculated [11]. Experimental values are given in [12] and [13] for the eclipse corona observed two years before solar minimum (1973), and at sunspot maximum (1980). Strong dispersion is plainly visible. Beyond a distance of three solar radii, neither theoretical nor measured values are reliable. Hydrostatic equilibrium no longer applies, and the predominance of the F corona interferes with measurement of P_K, because of background 'noise'.

This was also the case at distances less than three solar radii, within the polar coronal hole of 1973, when a special analysis was carried out [2] with a view to determining N_e as a function of radial distance. However, a coronal hole is homogeneous neither in density, nor in temperature, with plumes and small jets which appear clearly on a beautiful computer-processed photograph of the eclipse [2, 4]. This problem is a long way from being resolved.

The most up-to-date model of a coronal hole, based upon ATM/Skylab data, differs from measurements taken during eclipses [2, 22], and does not take inhomogeneities into account. It is hoped that the space-borne Lasco externally-occulted coronagraphs [25] aboard SOHO will carry out new and more accurate measurements, especially of the well-known polar plumes.

Finally, there is the particular case of the great streamers studied during eclipses, together with their polarisation [14]. The contribution of a streamer near the plane of the limb ($\theta = 90°$) can be measured as a modulation Δl of the azimuthal distribution of intensities. When measuring the corresponding degree of polarisation, it is also possible to evaluate θ when the geometrical contribution ΔI along the line of sight is not too great; that is, if the streamer is geometrically 'thin'. It then follows that equation (B.8) can be rewritten in a simplified form:

$$\Delta I\ (\rho) = B_{\odot} \,.\, \sigma_0 \,.\, W_\lambda(r) \,.\, N_e\ (r) \,.\, \Delta l \tag{B.9}$$

A complete analysis of such a case appears in [14]. There, distribution of N_e in the section of the streamer is given by an approximate formula:

$$N_e = C^{ste} \,.\, \exp - [(x^2 + \eta \,.\, y^2) \,.\, \textstyle\sum^{-2}] \,.\, \exp\ (\gamma \,.\, z) \tag{B.10}$$

where η is the ratio of extent along the line of sight and observed thickness in the plane of the image; and $\sum$ is the corresponding geometrical section of streamer.

If a large streamer is composed of several overlapping parts (sheets), it is impossible to consider the distribution in detail of N_e, but a section in the form of a Gaussian curve is taken as a good first approximation. Figure B.4 shows radial variations of N_e obtained for different cases processed from eclipse images.

A remarkable change in the behaviour of N_e is observed between values of 3 and 3.5 solar radii, which correspond to the lowest values of the effective section $\sum$. Assuming that this part corresponds to a velocity v_s of rapid expansion of the gas crossing the sound barrier 'within' the streamer (by analogy with a Laval nozzle), and using the equation of conservation of mass

$$N_e\ (r) \,.\, \textstyle\sum(r) \,.\, v(r) = C^{ste} \tag{B.11}$$

$v(r)$ can then be calculated. An increasing velocity is obtained, out to the limit of visibility of the streamer. Consequently, at present levels of accuracy, curvature is neither predicted nor observed. This procedure has been applied more recently to a longer and narrower streamer of the 1973 eclipse, and velocities obtained are given in [24]. Again, no curvature of the streamer was observed. It is evident, though, that velocities deduced decrease beyond five solar radii, and, taking into account the angular position of the streamer in space, a curvature would have been observed if the gas had been expelled unopposed, and if the magnetic field had been 'frozen in' within this gas.

This is evidently not the case; a modification of the model [6, 8] seems to be needed which would assume an outflow of azimuthal electric currents from either side of the streamer, beyond three solar radii, in such a way as to produce a magnetic force capable of confining the plasma. This is, though, quite another story [30].

ANALYSIS OF FINE STRUCTURES

Introduction

Although the inhomogeneous nature of the solar corona, even when quiet, has long been known, no concrete description of it has been produced, either in the theory of the acceleration of the solar wind, or in theories of the extended heating of the corona. Only mean values are considered, even in numerical simulations.

Theoretical work is thus founded essentially upon values which fail to take account of spatial and temporal variations. Sometimes, a certain 'turbulence' of the plasma is mentioned. A good photograph of the eclipse corona [2, 4] illustrates well, however, that coronal plasma is confined within fine structures (Figures B.5

and B.8). These structures join up, moreover, to form much larger structures, such as a streamer. So we can state that the corona is streaked with fine, elongated structures, as predicted by H. Alfvén [19]. All the evidence points to the fact that the physics of fine structures is different, for now we are dealing with a completely 'magnetised' plasma – where mean free paths are very long compared with radii of gyration around magnetic field lines [20]. Taking into account the filling factor, densities increase in a striking way and their transverse gradients reach high values, which suggests discontinuities. Structures may also be detached or twisted. Lastly, fine structures can interact and cross over each other, which introduces the possibility of reconstruction/destruction phenomena occurring on even smaller scales. Note that the finest coronal structure yet observed [27] is only a fraction of an arcsec across.

The study of these fine structures even today remains a frustrating pursuit, for various reasons:

- The extent of small-scale inhomogeneities is unknown. Also, no statistical data have been obtained because of the effects of geometrical superposition, although the (often unresolved) structures tend to extend radially. They are elongated in the direction of the magnetic field, which is stretched out by the solar wind as distance increases.
- The lack of a general approach to this problem hinders any linkage with the more general problem of stellar coronae and stellar winds. Until now, attention has been focused only on large streamers and coronal holes [8, 15];
- Optical images, which might potentially give the best resolutions, are often obtained with instruments of inadequate aperture and 'noisy' detectors, and results are therefore disappointing. In addition, modern detectors such as CCD cameras are still not commonly in use during eclipses, and it is not at the moment certain that coronal imagery would benefit from their use.

The total eclipse of 11 July 1991 was, however, an exception, because of the unique opportunity of using the world's best optical telescopes and detection systems, on Mauna Kea (Hawaii). Some very interesting figures are available, which confirm the existence of very fine structures in the corona [20, 27, 29]. There is no doubt that these observations will be improved upon in the future, with the help of ever more sophisticated detectors and better optics, on portable telescopes.

Sheets and discontinuities

On a good photograph of the corona in white light taken during an eclipse, several quasi-radial sheets, slightly curved, can be identified. They often form part of a large coronal streamer. On almost every occasion, they show on one side the characteristic signature of a very strong transverse gradient (a kind of discontinuity): a large and definite 'jump' in N_e, when traversing this doubtless convex 'side' [16]. These N_e jumps can be likened approximately to a tangential discontinuity, as it is defined in plasma physics, such is the value of the transverse gradient, surpassing even the strictly radial gradient determined by gravity (equation (B.6)), or, possibly, the dynamic pressure gradient associated with the solar wind $\nabla(\frac{1}{2}\ NM_H v^2)$ of velocity

Table B.1. Parameters of coronal plasma

r $[R_\odot]$	1.5	2.0	2.4
N $[e^- . cm^{-3}]$	1.1×10^8	1.7×10^7	0.4×10^7
δl [Mm]	4.0	4.5	5.0
$\nabla_\perp N$ $[e^- . cm^{-4}]$	0.25	0.027	0.006
$\nabla_r N$ $[e^- . cm^{-3}]$	0.77×10^{-3}	10^{-4}	4×10^{-5}
ratio $\nabla_\perp N/\nabla_r N$	>325	>270	>150
$\delta B_\odot$ [gauss]	0.8	0.25	0.12
T_{hyd} $[10^6$ K]	1.25	1.0	0.8
$R_L = mvc/eB_l$ [cm]	14	40	80

$v = v(r)$. Let us remind ourselves of the characteristics of a coronal discontinuity D, obtained through analysis of a typical case of a coronal sheet of known radius of curvature [16] in longitude and latitude (although note that we are not considering a perfectly 'flat' sheet).

Table B.1 gives typical values for the parameters of coronal plasma deduced for the dense edge of such a coronal sheet [16]. δl is the observed 'width' of the discontinuity. It is close to the value of the spatial resolution of the image. The density jump could be in reality greater. However, the deduced figures are already very significant. The density gradient $\nabla_\perp N$ through D is greater by several orders of magnitude than the radial gradients! Often a 'darkening', or vacuity, seems to exist along D, which makes the jump even greater. The strength of the magnetic field has been calculated from the conditions needed to obtain a stable D. This stability is confirmed – at least for the mid-corona – by comparing eclipse photographs taken several hours apart. This problem has been studied in greater detail in a more theoretical paper [28], which also examines dynamics. It is probable, though, that the origin of these discontinuities should be sought largely in the relatively violent phenomena 'restructuring' the corona after the ascent of a coronal mass ejection.

Let us assume a constant temperature in D. We note that the radius of gyration of the electrons $R_L << \delta l$ – which signifies that the plasma is completely magnetized and that the evolution of D – is slow. If N_1 and N_2 are densities ($N_1 >> N_2$) at either side of D, $\vec{v}$ is the local velocity and $\vec{B}$ the magnetic field, we have equations for the different components:

$$\begin{aligned} &v_\perp \equiv 0 \\ &v_{||1} \neq v_{||2} \\ &B_\perp \equiv 0 \\ &B_{||1} \neq B_{||2} \\ &N_1 \neq N_2 \end{aligned} \qquad \text{(B.12)}$$

The current $j_\perp$ across D is similarly zero because there is no flux of particles across D, but a displacement of the whole of D is not ruled out.

If P_1 and P_2 are the gas pressures at either side, we finally have

$$P_1 + B^2{}_1/8\pi = P_2 + B^2{}_2/8\pi \qquad \text{(B.13)}$$

Taking $B_1 \approx 0$, $P_j = 2N_{ej} . k . T_{hyd}$, we obtain the value of B_2 ($B_2 = \delta B_\perp$) for Table B.1 – no doubt the best estimate of the true value of the magnetic field in the mid-corona.

A more sophisticated analysis would involve the equation $\vec{j} = {}^{c}/_{4\pi}\ \vec{\nabla} \wedge \vec{B}$, to calculate the magnetic force (Ampère) and the processes at the interface, if the level of current is significant. Also, interaction with MHD waves propagating through the corona is certainly fundamental [28].

Surges, fibrils and plasmoids

These small structures of the inner corona and mid-corona can be observed only on eclipse photographs of very high spatial resolution. They seem to be naturally associated with violent and localised coronal phenomena in the transition region, which occur all over the Sun: above tiny regions of ephemeral activity which doubtless lie near nodes of the chromospheric network (although phenomena outside the network cannot be excluded); above large chromospheric filaments (or prominences); or more especially above active regions with multiple sunspots; or, finally, above the polar regions (fine structures, plumes or rays).

The first reports of visual and photographic detection of very fine coronal jets within the corona (called 'jetlets', or 'spikes') came from Japanese observers, and Russian groups in Moscow and Kiev (Ukraine). The first photometric studies were carried out in 1968–73 [17], and since then, new measurements have been obtained [18] which confirm the first theoretical predictions [19] of filamentation in the coronal plasma. The main factor involved is the transverse photometric 'cross-section' (or total width at mid-height) which should be comparable to the (unknown) extent along the line of sight. Then, using equation (B.9), N_e can be obtained.

From our estimation of the transverse section $\sum$, with $\Delta l \approx$ total width at mid-height, and using equation (B.11), we obtain the variations of velocity $v(r)$. But more reliable values can be derived from observation of the proper motion. Few measurements exist, however, given the difficulty inherent in measuring proper motion or interpreting results of profile analysis of emission lines, since Doppler velocities (along the line of sight) are needed to obtain all the components of the velocity vector [17]. Experiments carried out during the 1991 eclipse, with the resolving power of the CFH Telescope, provided new results [27].

Somewhat surprisingly, fine structures less than one arcsec across were seen as very time-variable. Elongated coronal filaments less than 10^6 m (1 Mm) in diameter were certainly detected above active regions, at more then 100 Mm from the surface, with lifetimes of the order of 100 seconds (Figure B.8).

Fibre-like structures of greater diameter have longer lifetimes. The proper motions of the fibrils are apparent, but seem to vanish when the coronal background changes.

A quite different phenomenon was observed with the CFH Telescope: an isolated cloud of coronal plasma, or plasmoid, 'dissolving' into the surrounding corona and associated with numerous dynamical phenomena [20, 27]. Accurate measurements of the position of the cloud's centre of gravity were taken, in a sequence of more than 6,000 CCD video frames. However, only the component of the proper motion

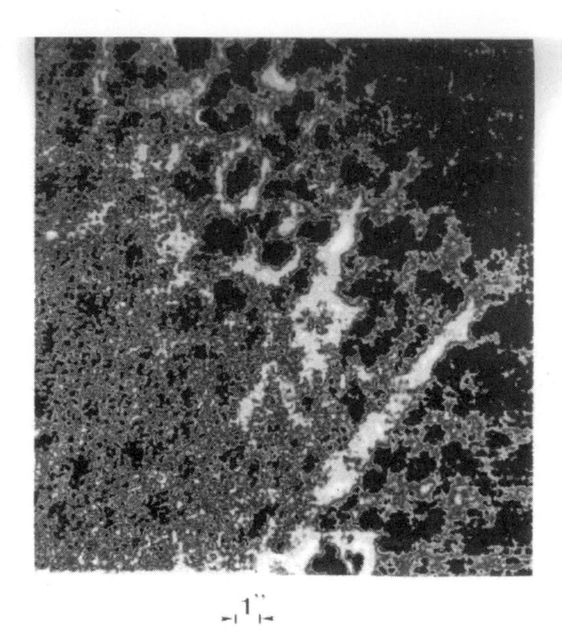

Fig. B.8. The smallest coronal structure detected during high-resolution observations with the CFH Telescope during the eclipse of 11 July 1991, in white light. The lifetime of this coronal fibril is about one minute. Its true diameter, corrected for atmospheric spreading, is less than 300 km, and its temperature is in the region of 1–2 million K. (IAP/CNRS.)

projected onto the limb plane can be derived. Also, the cloud, or plasmoid, about three arcsec across at the beginning of the video sequence, splits several times, producing filamentary structures or short-lived fibrils. Finally, after 200 seconds, the cloud disappears. Its proper motion v_{tot} against the plane of the sky is of the order of 100 km s^{-1}. This observation of a coronal plasmoid, and its associated dynamical phenomena, requires more thorough analysis [20], but suggests that a large part of the corona – if not the entire corona – is in a dynamic state at a resolution of less than 1 arcsec. Ejected plasma clouds are partly diamagnetic [21], which means that the external field does not 'penetrate' them, and since we see them disappearing against the coronal background it may be that small-scale MHD phenomena are in action, and could even be responsible for the elusive mechanisms behind coronal heating. We have here a whole new field of coronal observations made during eclipses, calling for spatial resolutions down to at least 1 arcsec. Nevertheless, plasmoids on much larger scales, associated with vast outward leaps of 'coronal' prominences, may be observed. During a total eclipse, a great deal of the corona is observed, and when the Sun is active the possibility of the amateur being able to observe this phenomenon is by no means ruled out.

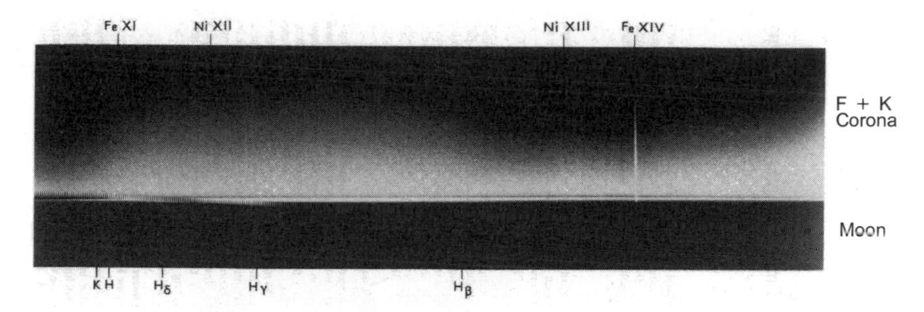

Fig. B.9. This photograph shows one of the coronal spectra obtained during the eclipse of 30 June 1973 from Chad, by the team from the Institut d'Astrophysique de Paris/CNRS. The slit of the spectrograph was aligned radially in the vicinity of the limb, and covered a distance of two solar radii from the Moon. The film-duplicate registering the edge of the Moon was exposed for ten times longer than that covering the corona, in order to bring out very weak parasitic lines of chromospheric origin. (G. Stellmacher and S. Koutchmy, IAP/CNRS.)

TEMPERATURES IN THE CORONA

An addition to the title of this section might be: '... not to mention sites of particle acceleration in the corona.' Different coronal temperatures may be investigated by observing forbidden coronal emission lines. These lines are produced in a medium much diluted by collisions and radiative excitations of strongly ionised atoms. Figure B9 shows a classic eclipse spectrum, revealing many lines, some of which are dealt with in Table B.2. The theory of the formation of these lines has been of great importance to fundamental atomic physics, one reason being its analysis of the equilibrium of ionisation of different ions.

Studies are currently being carried out on very small effects on these lines: for example, the Hanle effect, which produces linear polarisation of the lines, modified in the presence of a magnetic field. Nevertheless, in recent decades, studies have concentrated on the analysis of (allowed) lines in extreme UV and soft X-ray wavelengths. Note that the measurement of radiative flux F_i in a line is at once a non-linear function of the plasma density, proportional to electron density N_e, and of local ion temperature T_i:

$$F_i = \int N_e^{\alpha} \,.\, P(T_i) \,.\, dV \qquad \text{(B.14)}$$

where dV is an element of the emissing volume.

The coefficient α is equal to 2 for the allowed lines, and is slightly less than this value for forbidden lines. Function $P(T_i)$, known as radiative loss, is calculated for each line from analysis of the ionisation equilibrium. It has a well-defined maximum, and there is therefore a tendency to consider that a line is formed at the temperature where this function reaches its maximum. To obtain radiative flux F_i, one must evidently integrate for a volume of the corona, and generally on the basis of the element of spatial resolution integrated over the line of sight. Note that hydrostatic temperature, determined by the radial gradient of densities (see equations (B.6) and

(B.7)) is not very different from ion and electron temperatures (see below), when plasma flows are negligible compared with the thermal velocities of the particles. The notion of ion temperature can moreover be conveniently replaced by that of classes of lines, which are defined below. In fact, the emission line profile reflects this temperature only partially: to it must be added Doppler effects, as the corona is in a permanently dynamic state, especially on the small scale. If spatial resolution is sufficient to resolve coronal elements moving at different velocities along the line of sight where integration is being done, it becomes possible to measure the component of this velocity projected along the line of sight. As a general rule, this is not feasible, and an *ad hoc* velocity – the velocity of microturbulence v_m – is introduced. This is added to the thermal velocity of the ions, and can be evaluated at the same time as ion temperature T_i for a Maxwellian distribution of velocities. Of course, when collisions become rare – as happens in the mid-corona at about one solar radius from the surface – the distribution of velocities varies from that taken at thermal equilibrium. In the outer corona, mean free paths of ions become very long, and temperature becomes anisotropic. It decreases more rapidly in the radial direction of the solar wind, even if the magnetic field tends to preserve structures like the heliospheric sheet and magnetic sectors.

So, detailed study of the profile of coronal lines can help us to know velocities in the corona, and ion and electron temperatures. Electron temperature T_e is an interesting parameter *a priori*, since not only ought it to play a rôle in the equilibrium of ionisation (at thermal equilibrium, $T_i = T_e$ in a plasma), and in excitation of atomic levels through collision, it also determines conduction in the plasma, including the conduction of heat.

The determination of T_e is therefore very important when determining energy balances in the corona. Historically, it is of interest to note that the high temperature of the corona was hinted at long before the identification of the forbidden lines. For example, Grotrian, in 1933, was able to predict an electron temperature of nearly 10^6 K by evaluating the smearing of the F spectrum reflected by electrons (Thomson scattering); and Lyot, at about the same time, was evaluating the ion temperature (without knowing the exact mass of these ions!) producing coronal lines, by interpreting the measured width of these lines (about 0.09 nm for the green line) as being due to thermal agitation. Nowadays, such measurements may be carried out under much better conditions during eclipses, and their interpretation can fundamentally refine coronal physics, making especially possible: measurement of magneto-acoustic waves through Doppler effects; analysis of the radial variation of ion temperatures, and especially electron temperatures, for different typical regions at the equator, and at the poles in coronal holes.

These two types of measurement require very different methods, as spectro-spatial-temporal resolution and photometric quality of measurements differ greatly. In both cases, monochromatic imaging may be contemplated to obtain two-dimensional, simultaneous coverage. A solution in both cases involves the use of a Fabry–Pérot etalon. We believe that nothing can replace the analysis of good spectra, and this is the solution we advocate for future eclipses. Table B.2 shows the choice of lines.

It is now well established that there are three main classes of coronal emission line:

- Class I, where T_i = 1 to 1.5 MK, typical lines being those of FeIX, FeX (the well-known red line at 637.4 nm), and FeXI.
- Class II, where T_i = 1.5 to 2.5 MK, with typical lines FeXIV (the well-known green line at 530.3 nm), FeXIII, and FeXV.
- Class III, formed at T_i > 3MK, which exist only in very active regions and eruptions. Typical here is the line of CaXV, in the yellow.

Table B.2. Forbidden lines of the corona, in the visible and near-infrared (CCD) (after Jefferies and colleagues, 1971 [31]). Note that intensities are given in relation to the adjacent continuum of width 0.1 nm, and for typical regions of the corona.

λ (nm) wavelength	Transition ion, state of ionisation and line transition	$A(s^{-1})$ probability of de-excitation	Potential excitation (eV)	Potential ionisation (eV)	Intensity (inner corona) relative units
332.8	CaXII $2p^5\ ^2P_{1/2} \rightarrow {}^2P_{3/2}$	488	3.72	589	(17)
338.8	FeXIII $3p^2\ ^1D_2 \rightarrow {}^3P_2$	87	5.96	325	37
360.1	NiXVI $3p\ ^2P_{3/2} \rightarrow {}^2P_{1/2}$	193	3.44	455	(18)
364.29	NiXIII $3p^4\ ^1D_2 \rightarrow {}^3P_1$	18	5.82	350	1.5
398.69	FeXI $3p^4\ ^1D_2 \rightarrow {}^3P_1$	9.5	4.68	261	
408.63	CaXIII $2p^4\ ^3P_1 \rightarrow {}^3P_2$	319	3.03	655	(22)
423.14	NiXII $3p^5\ ^2P_{1/2} \rightarrow {}^2P_{3/2}$	23	2.93	318	8
441.2	ArXIV $2p\ ^2P_{3/2} \rightarrow {}^2P_{1/2}$	112	2.84	682	16
511.603	NiXIII $3p^4\ ^3P_1 \rightarrow {}^3P_2$	157	2.42	350	2
530.286	FeXIV $3p\ ^2P_{3/2} \rightarrow {}^2P_{1/2}$	60	2.34	355	190
544.5	CaXV $2p^2\ ^3P_2 \rightarrow {}^3P_0$	83	4.45	814	(15)
553.9	ArX $2p^5\ ^2P_{1/2} \rightarrow {}^2P_{3/2}$	106	2.24	421	5
569.442	CaXV $2p^2\ ^3P_1 \rightarrow {}^3P_0$	95	2.18	814	(28)
637.451	FeX $3p^5\ ^2P_{1/2} \rightarrow {}^2P_{3/2}$	69	1.94	233	40
670.183	NiXV $3p^2\ ^3P_1 \rightarrow {}^3P_0$	57	1.85	422	(27)
705.962	FeXV $3s3p\ ^3P_2 \rightarrow {}^3P_1$	38	31.77	390	5
789.194	FeXI $3p^4\ ^3P_1 \rightarrow {}^3P_1$	44	1.57	261	50
802.421	NiXV $3p^2\ ^3P_2 \rightarrow {}^3P_1$	22	3.39	422	
1074.680	FeXIII $3p^2\ ^3P_1 \rightarrow {}^3P_0$	14	1.15	324	100
1079.795	FeXIII $3p^2\ ^3P_2 \rightarrow {}^3P_1$	9.7	2.30	325	50

These temperatures (T_i) are essentially calculated on the basis of analysis of equilibrium of ionisation, and are attributed to ions. Black-and-white photographs of emissions, observed during the 1981 eclipse, in the green line of FeXIV and the red line of FeX, and processed using the *Mad Max* algorithm to increase visibility of structures (Figure B.10), show a lack of any obvious correlation between these different, though simultaneous, emissions.

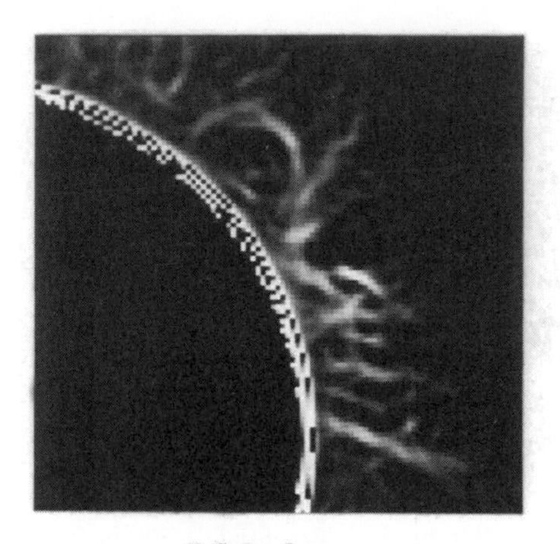

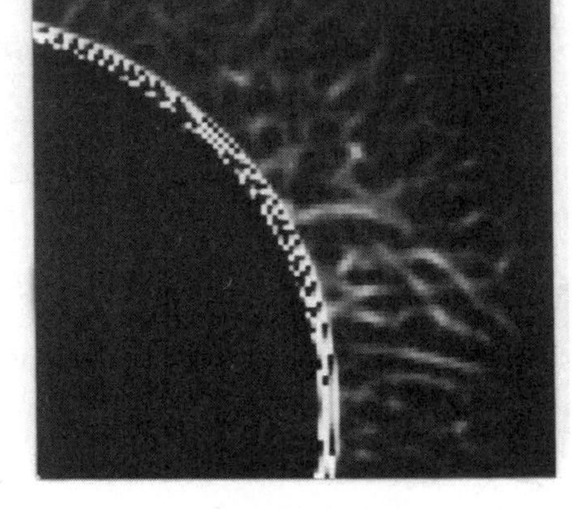

Fig. B.10. Images processed with the algorithm *Mad Max*, showing the same part of the corona during the eclipse of 16 February 1980, observed in two very different coronal lines ($T = 10^6$ K, and $T = 2 \times 10^6$ K), with narrow interferential filters. (J. Sykora, Slovakian Academy of Science; digital treatment by O. Koutchmy, Paris VI, and IAP/CNRS.)

There are therefore great inhomogeneities in temperatures. Looking at photographs taken at the same time in white light, showing plasma density distribution, we have noticed a correlation with the FeXIV image ($T = 2$ MK), which seems to correspond with the most likely coronal temperature, outside centres of activity. This result was already apparent when the so-called hydrostatic temperature (see equations (B.6) and (B.7)) was calculated for the main feature of the inner corona (Figure B.4).

We conclude from this that the mean temperature of the fairly dense inner corona is about 1.5–2 MK, varying by no more than a factor of two, though producing important effects on consecutive emissions of the different ions. However, local gas pressure is not particularly affected by these changes in temperature: firstly, because inhomogeneities in density are much more significant; and secondly, because flows must also be considered. This is undoubtedly not the case with coronal eruptions observed at soft X-ray wavelengths, where energetic, if intermittent, processes are at work. And all this gives us no information about temperatures in coronal holes, which produce very little ion emission. Where high temperatures of coronal condensations and the even higher temperatures in the heart of flares are concerned, visible lines are not of much use. Only X-ray investigations can assess temperatures in flares, which reach tens of millions of degrees.

CONCLUSIONS

Future eclipse observations should provide useful support for space missions such as Yohkoh, and especially SOHO, as well as paving the way for the solar probe mission which will send one or more probes to observe the corona from nearby and within.

Since photometry carried out during eclipses is very accurate – even excellent when field stars are used for photometric calibration – an absolute standard for observations from space may soon be available. Wide-field imaging with good resolution requires the use of a neutral-density radial-gradient filter.

If eclipse teams at well-separated sites collaborate, it is possible to take advantage of the rigid rotation of the corona to obtain stereoscopic images of great value [23].

In the future, eclipse observations could attempt measurements not easy to secure from space platforms: for example, analysis of the profile of emission lines of the mid-corona, and infrared analysis, including polarimetry. This is necessary if progress is to be made in our understanding of the rôle of the magnetic field, accessible through measurement of the polarisation of emission lines; and in our ability to detect MHD waves, using the Doppler effect on these lines in time sequences. It is to be hoped, moreover, that amateur eclipse observers of the future will contribute more to the study of small-scale phenomena, such as the plasmoid observed in 1991. A moderate-aperture telescope (at least 150 mm) will be needed, but it is necessary to be able to go down to details smaller than one arcsec, which is theoretically possible by taking advantage of very short exposures and the *post facto* accumulation of numerous images taken in rapid sequence.

White-light observations from orbiting space vehicles – free from image-degrading factors such as atmospheric disturbance – could be more effective in approaching this problem, but as yet this is but a dream. A ground-based coronagraph with large mirrors for near-infrared work would be able to make such an observation when the Sun is not eclipsed. It should involve a simple adaptive-optics system. Such an instrument in orbit would enable ultimate analysis of short-lived structures of dimensions less than 1 arcsec, which rank at present among the most mysterious turbulent magnetic phenomena.

REFERENCES

1. L. Golub and J. Pasachoff, *The Solar Corona*, Cambridge University Press, 1997.
2. a. S. Koutchmy, Study of the June 30, 1973 Trans-Polar Coronal Hole, *Solar Physics*, **51**, 399, 1975.
   b. C. Lebecq, S. Koutchmy and G. Stellmacher, The 1981 Solar Eclipse: II. Global Absolute Photometrical Analysis, *Astronomy and Astrophysics*, **52**, 157, 1985.
3. M. Loucif and S. Koutchmy, Solar Cycle Variations of Coronal Structures, *Astronomy and Astrophysics Supplement*, **77**, 45, 1989.
4. S. Koutchmy, Solar Corona, in: *Illustrated Glossary for Solar and Solar-Terrestrial Physics, Eds.* Bruzek and Durrant, Reidel, 1977.
5. a. K. Schatten, Large-Scale configuration of the Coronal and Interplanetary Magnetic Fields, Ph. D., Berkeley, 1968.
   b. M.D. Altschuler and G. Newkirk, *Solar Physics*, **9**, 131, 1969.
   c. P. Ambroz, The Coronal Magnetic Field – Numerical Expansion Methods and

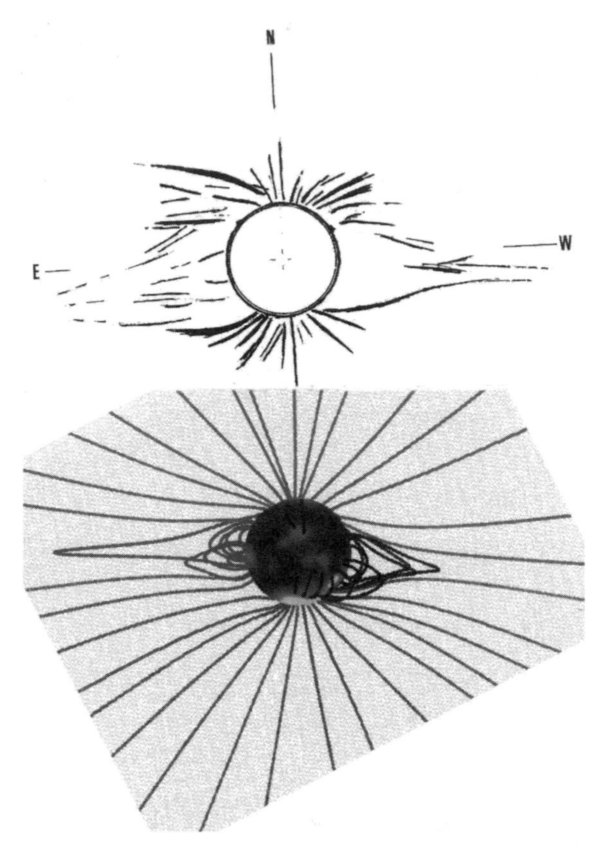

Fig. B.11. Top: a summary sketch of coronal structures, from photographic observations by M. Molodensky's team of the eclipse of 9 March 1997 in Siberia (1 h 10 m UT). Bottom: 3-D representation of lines of force calculated before the eclipse by American scientists Jon Linker and Zoran Mikic (Science Applications International Corporation of San Diego), to model the solar corona with a view to prediction. This work is extremely valuable in verifying modelling techniques and in our understanding of coronal phenomena. Note that scales and times are the same for both observation and calculations.

Observations, in: *Solar Magnetic Fields and Corona*, Vol. 1, Proceedings XIIIth Consultation Meeting on Solar Physics, Nauka, Sib. Div., 1989.
6. T.J. Hoeksema, Extending the Sun's Magnetic Field through the Three-Dimensional Heliosphere, *Advanced Space Research*, vol. 9, 1989.
7. B.V. Somov and S.I. Syrovatsky, Appearance of a Current neutral. Sheet in a Plasma Moving in the Field of a 2 D Magnetic Dipole, *J.E.T.F.*, **61**, 1971.
8. S. Koutchmy and M. Livshits, Coronal Streamers, *Space Science Review*, **61**, 393, 1992.
9. a. J. Heyvaerts et coll., Aspects Magnétohydrodynamiques de l'activité solaire, *Annales de Physique*, vol. 5–6, 1980.
   b. E. Priest, Solar MHD. Riedel, 1982.

10. S. Koutchmy and P. Lamy, The F-Corona and the Circum-solar Dust: Evidences and Properties, in: *Proceedings IAU* Coll. No 85, Giese and Lamy Eds, Reidel, **63**, 1985.
11. K. Saito, A Non-Spherical Axisymetric Model of the Solar K-Corona of the Minimum Type, *Annals of the Tokyo Astronomical Observatory*, **XII**, 5, 1972.
12. S. Koutchmy, J.-P. Picat and M. Dantel, Polarimetric Analysis of the Solar Corona, *Astronomy and Astrophysics*, **59**, 349, 1977.
13. J. Dürst, Two Colour Photometry and Polarimetry of the Solar Corona of 16 February 1980, *Astronomy and Astrophysics*, **112**, 241, 1982.
14. S. Koutchmy, Étude hydrodynamique du grand jet coronal observé à l'éclipse du 7 mars 1970, *Solar Physics*, **24**, 374, 1972.
15. J.B. Zirker, *Coronal Holes and High Speed Wind Streams*, Colorado Association University Press 1977.
16. S. Koutchmy, Un modèle de grand jet coronal avec renforcement de région active, *Astronomy Astrophysics*, **13**, 79, 1971.
17. a. G. Stellmacher and S. Koutchmy, Study of Low Dispersion Eclipse Spectra, *Astronomy and Astrophyics*, **35**, 43, 1974.
    b. S. Koutchmy and G. Stellmacher, Photometric Study of Chromospheric and Coronal Spikes Observed during the Total Eclipse of 30 June 1973, *Solar Physics*, **49**, 253 1976.
18. S. Koutchmy, Small-scale Coronal Structures, in: *Solar and Stellar Coronal Structure and Dynamics*, Proceedings of the NSO/SP 9th Workshop, R.C. Altrock *Ed.*, 1988.
19. H. Alfvén, On the Filamentary Structure of the Solar Corona, in: *The Solar Corona, IAU Symp. 16*, Evans *Ed.*, Academic Press, 1963.
20. a. J.C. Vial, S. Koutchmy and the CFHT, Team, Evidence of Plasmoid Ejection in the Corona from 1991 Eclipse Observations with the CFHT, in: *Proceedings of an ESA Workshop on Solar Physics and Astrophysics at Interferometry Resolution*, 1992.
    b. C. Delannée, S. Koutchmy, I. Veselovsky and A. Zhukov, Coronal Plasmoids: MHD parameters, *Astronomy and Astrophysics*, **329**, 1111, 1998.
21. G.W. Pneuman, Ejection of Magnetic Fields from the Sun: Acceleration of a Solar Wind containing Diamagnetic Plasmoids, *Astrophysical Journal*, **265**, 468, 1983.
22. F. Crifo-Magnant and J.-P. Picat, A Density Model for the N-Polar Coronal Hole at the 1973 Eclipse, *Astronomy and Astrophysics*, **88**, 97, 1980.
23. S. Koutchmy and M. Molodensky, Three-Dimensional image of the Solar Corona W-L observations of the 1991 Eclipse, *Nature*, **360**, 717, 1992.
24. S. Koutchmy, Streamer Eclipse Observations, in '1st SOHO Workshop', Annapolis, August 25–28, 1992, ESA SP, 1992.
25. S. Koutchmy, Spaceborne Coronography, *Space Science Review*, **47**, 95, 1988.
26. Ph. Lamy, J.R. Kuhn, H. Lin, S. Koutchmy and R.N. Smartt, No Evidence of a Circumsolar Dust Ring from IR Observations, *Science*, **257**, 1994.
27. S. Koutchmy and the CFHT, Eclipse Observations of the very fine-scale Solar Corona, *Astronomy and Astrophysics*, **281**, 249, 1994.

28. S. Grib, S. Koutchmy and V.N. Sazonova, Some MHD interactions in coronal structures, *Solar Physics*, **169**, 151, 1966.
29. L. November and S. Koutchmy, Coronal W-L dark loops and density fine structure, *Ap. J.*, **466**, p. 512–528, 1996.
30. S. Koutchmy, M. Molodensky and D. Vibert, *IAU 167, Proceedings*, 1997.
31. J.T. Jefferies, F.Q. Orall and J.B. Zirker, The spectrum of the inner corona observed during total eclipses, *Solar Physics*, **16**, 103, 1971.

Appendix C

Computer program for lunar and solar eclipse dates

This program, written in standard BASIC, is easily transposed to other computer languages.

Meaning of certain variables

YR	Target year (Input data)
DY	Day
S	Sun's geocentric longitude
M	Moon's geocentric longitude
N	Geocentric longitude of Moon's ascending node
MLM	Mean longitude of Moon
MLS	Mean longitude of Sun
MLA	Mean lunar anomaly
MSA	Mean solar anomaly
CLM	Celestial longitude of Moon
JC	Julian century
JM	Julian millennium
DL	Difference in longitudes of Sun and Moon
DN	Difference in longitudes of Sun and ascending node

```
10    REM "Calculation of dates of lunar and solar eclipses"
20    REM "Introduction of data (selected year)"
30    INPUT "Enter selected year"; YR
40    REM "Initialisation"
50    DY=1
60    B=-360
70    C=-360
80    PI=3.1415926535
90    J=DY-1
100   REM "Millennia and centuries from 2000.0"
110   AB=YR/4-INT(YR/4)
120   IF AB=0 THEN AB=1
```

```
130    JM=JC/10
150    REM "Calculation of mean elements"
160    MLS=4.895062967+6283.319663*JM+5.300181*JM*JM
170    KAS=0.003740816-0.004793106*JM+0.000281128*JM*JM
180    HAS=0.016284477-0.001532379*JM-0.000720171*JM*JM
190    LPS=ATN (HAS/KAS)
200    ES=ABS (HAS/SIN(LPS))
210    MAS=MLS-LPS
220    AES=MSA
230    FOR I=1 TO 5
240    AES=MSA+ES*SIN (AES)
250    NEXT I
260    AVS=2*ATN (SQR ((1+ES)/(1-ES))*TAN (AES/2))
270    ARS=LPS + AVS
280    LCS=ARS
290    IF LCS<0 THEN LCS+2*PI
300    LCS=2*PI*(LCS/2/PI-INT (LCS/2/PI))
310    P=5.19846674+7771.377144*JS-2.8509468E-5*JC*JC
320    Q=1.627905233+8433.466157*JC-5.9453671E-5*JC*JC
330    MLA=2.355555898+8328.691424*JC+1.5696909E-4*JC*JC
340    MLM=3.810344426+8399.709112*JC-2.3209485E-5*JC*JC
350    N=MLM-Q
360    CLM=MLM + 0.10975981*SIN (MLA)+3.728314E-3*SIN (2*MLA)
370    CLM=CLM+1.751147E-4*SIN (3*MLA)+0.0114895994*SIN (2*P)
380    CLM=CLM-3.2390886E-3*SIN (MSA)-6.0674432E-4*SIN (P)
390    CLM=CLM+0.02223564*SIN (2*P-MLA)-1.9955415E-3*SIN (2*Q)
400    CLM=CLM+1.0261566E-3*SIN (2*P-2*MLA)+9.9852225E-4*SIN
       (2*P-MLA-MSA)
410    CLM=CLM+9.3059986E-4*SIN (2*P+MLA)+8.0066979E-4*SIN
       (2*P-MSA)
420    CLM=CLM+7.1602132E-4*SIN (MLA-MSA)-5.3169516E-4*SIN
       (MLA+MSA)
430    CLM=CLM+2.674717E-4*SIN (2*P-2*Q)-2.1865097E-4*SIN
       (2*Q+MLA)
440    CLM=CLM-1.91664684E-4*SIN (2*Q-MLA)+1.8631389E-4*SIN (4*P-
       MLA)
450    CLM=CLM+1.4917716E-4*SIN (4*P-2*MLA)-1.3802645E-4*SIN
       (2*P-MLA+MSA)
460    CLM=CLM-1.183915E-4*SIN (2*P+MLA)-9.0223826E-5*SIN (P-
       MLA)
470    CLM=CLM+8.7363425E-5*SIN (P+MLA)+7.0685834E-5*SIN
       (2*P+MLA-MSA)
480    CLM=CLM+6.9764688E-5*SIN (2*P+2*MLA)+6.7389101E-5*SIN
       (4*P)
490    CLM=CLM+6.394692E-5*SIN (2*P-3*MLA)
```

```
500    CLM=2*PI*(CLM/2/PI-INT (CLM/2/PI))
510    IF CLM<0 THEN CLM=CLM+2*PI
520    N=2*PI*(N/2/PI-INT (N/2/PI))
530    REM "calculation of DL and DN for solar eclipse"
540    DL=CLM-LCS+2*PI
550    DL=2*PI*(DL/2/PI-INT (DL/2/PI))
560    DN=ABS (LCS-N)
570    DN=2*PI*(DN/2/PI-INT (DN/2/PI))
580    REM "Verification of conditions for DN and DL for solar eclipse"
590    REM "Calculation of time"
600    IF ABS (DL*180/PI-B*180/PI)>15 AND 180/PI*DN<11 THEN
       D=B*180/PI : H=24*(360-D)/(DL*180/PI-D) : PRINT 'SOLAR
       ECLIPSE' : PRINT 'Day : ';DY-1 : PRINT 'Hour' ;INT (100*H+0.5)/
       100
610    IF ABS (DL*180/PI-B*180/PI)>15 AND ABS (180/PI*DN-180)<11
       THEN D=B*180/PI : H=24*(360-D)/(DL*180/PI+360-D) : PRINT
       'SOLAR ECLIPSE' : PRINT 'Day : ';DY-1 : PRINT 'Hour' ; INT
       (100*H+0.5)/100
620    B=DL
630    REM "Calculation of DL and DN for lunar eclipse"
640    DL=CLM-LCS+3*PI
650    DL=2*PI*(DL/2/PI-INT (DL/2/PI))
660    REM "Verification of conditions for DN and DL for lunar eclipse"
670    REM "Calculation of time"
680    IF ABS (DL*180/PI-C*180/PI)>15 AND 180/PI*DN<15 THEN
       D=C*180/PI : H=24*(360-D)/(DL*180/PI+360-D) : PRINT 'LUNAR
       ECLIPSE' : PRINT 'DAY : ';DY-1 : PRINT 'Hour' ;INT (100*H + 0.5)/
       100
690    IF ABS (DL*180/PI-C*180/PI)>15 AND ABS (180/PI*DN-180)<15
       THEN D=C*180/PI : H=24*(360-D)/DL*180/PI+360-D) : PRINT
       'LUNAR ECLIPSE' : PRINT 'Day : ';DY-1 : PRINT 'Hour';INT (100*H
       + 0.5)/100
700    REM "Initialisation and incrementation"
710    C = DL
720    DY = DY + 1
730    REM "New calculation cycle or end of year sought"
740    IF DY>365 THEN PRINT "End of year sought" ;YR : STOP
750    GO TO 90
760    END
```

Appendix D

The eclipse of 11 August 1999

On 29 June 1927, the shadow of the Moon crossed mainland Britain during a total eclipse of the Sun – for the first time since 1724. Most observers along the track, in North Wales and northern England, did not see the eclipse because of poor weather; but now there is another chance to see such an event from England, and it may be a memorable occasion.

THE TRACK OF THE MOON'S SHADOW ON 11 AUGUST 1999

The last total eclipse of the millennium will begin 300 miles south of Nova Scotia (Canada), when the Moon's tapering shadow will touch down upon the Earth's surface at 09 h 01 m UT. At first, the shadowed zone will be 49 kilometres wide, and totality on the centre line will last for 47 seconds.

For 40 minutes the shadow crosses the North Atlantic, to make landfall on the Isles of Scilly at 10 h 10 m UT. It is now mid-morning, and the Sun stands 45° above the eastern horizon. Totality now lasts for 2 minutes for observers on the centre line, and the path of the shadow is 103 kilometres across. The shadow is moving at 910 m s^{-1}. One minute later, the shadow reaches Cornwall, crossing the peninsula in four minutes. Plymouth, in Devon, the largest UK town over which the shadow passes, is to the north of the centre line, and sees the Sun go dark for 1 minute 39 seconds. London does not experience totality, with 96.8% of the Sun hidden. At 10 h 16 m UT, the shadow leaves Britain and moves off into the English Channel. Jersey and Guernsey lie to the south of the shadow's path, and from there the Sun is more than 99.5% eclipsed.

The shadow then reaches France. Just as the northern part of the shadow leaves the English coast, its southern edge arrives on the Cotentin peninsula, at 10 h 16 m UT. Cherbourg is plunged into darkness for 1 minute 30 seconds. Four minutes later, the shadow reaches the *département* of Seine-Maritime, crossing Le Havre, Fécamp and Dieppe. It races across France, passing 30 kilometres to the north of Paris. The 'City of Light' sees the Sun, in the middle of the day at 10 h 23 m UT, 99.2% eclipsed, at an altitude of 51° above the horizon. Reims experiences the darkness of totality for 1 minute 59 seconds. Travelling ever eastwards, the shadowed zone clips the south of Belgium and Luxembourg. The streetlights of Metz will come on at 10 h 29 m UT during 2 minutes 13 seconds of totality. Strasbourg, to the north of the centre line, will enjoy 1 minute 22 seconds of totality.

As the shadow crosses Germany, Frankfurt sees a 97.9% partial eclipse, and from Stuttgart the eclipse is total for 2 minutes 17 seconds. At 10 h 35 m UT the Sun's altitude above the eastern horizon is 55°, and the shadowed zone is 109 kilometres wide, travelling at 740 m s^{-1}. Munich goes dark at 10 h 38 m UT, for 2 minutes 7 seconds.

At 10 h 41 m UT the shadow leaves Germany and enters Austria. The capital, Vienna, is 40 kilometres to the north of the shadow path, and from there the Sun is 99% eclipsed. The shadow touches the north-east of Slovenia, and passes into Hungary at 10 h 47 m UT. The shores of Lake Balaton are dark for 2 minutes 22 seconds, at 10 h 50 m UT. Like Vienna, Budapest is 40 kilometres off the track, and will see the Sun 99.1% eclipsed. As the shadow moves out of Hungary, its southern edge will brush across the north of Yugoslavia, before entering Romania.

The moment of maximum eclipse occurs at 11 h 03 m 04 s UT, when the axis of the Moon's shadow passes nearest to the centre of the Earth. At that moment, the centre of the shadow is in southern Romania, near the town of Ramnicu Valcea. The duration of totality is now at its greatest, at 2 minutes 23 seconds, with the Sun 59° above the horizon, and the 112-kilometre wide shadow is now travelling at 680 m s^{-1}. At 11 h 07 m UT, Bucharest, capital of Romania, right on the centre line, enjoys a totality of 2 minutes 22 seconds. The shadow moves off south-south-east, crossing into Bulgaria before sweeping into the Black Sea.

Reaching Turkey at 11 h 21 m UT, the shadow passes 150 kilometres north of the capital, Ankara – which sees a 96.9% eclipsed Sun – and travels on diagonally across the country. At 11 h 29 m UT, the town of Turhal is darkened for 2 minutes 15 seconds. The shadow crosses the border between Turkey and Iraq, as it also brushes the extreme north-east of Syria, at 11 h 45 m UT. At this time, at the centre of the shadow totality lasts for 2 minutes 5 seconds, and the Sun is 50° above the horizon. Baghdad, 220 kilometres to the south of the track, sees the Sun 94% eclipsed. The shadow enters Iran at 11 h 52 m UT, and for the next 30 minutes crosses the Iranian desert. Teheran lies to the north of the track, and sees a 94.3% eclipse. At 12 h 22 m UT the shadow arrives in Pakistan and runs alongside the Gulf of Oman. Nearby Karachi is in the shadow track, and is in darkness for 1 minute 13 seconds, with the Sun 22° high in the west. The shadowed zone is now only 85 kilometres wide. The shadow is moving at 2,000 m s^{-1}, and nearing the end of its journey it arrives at the last country on its track, India, at 12 h 28 m UT. As it crosses the subcontinent, the duration of totality falls below 1 minute. From Calcutta, the eclipse is partial, at 87.9%, with the setting Sun 2° above the western horizon. Passing north of Visakhapatnam, at 12 h 36 m UT, the shadow leaves India and plunges into the Bay of Bengal. It lifts off the surface of the Earth and sweeps into space at 12 h 36 min 23 s UT.

During its journey of 3 hours 7 minutes, the Moon's shadow has raced across nearly 14,000 kilometres, covering 0.2% of the Earth's surface.†

GENERAL CHARACTERISTICS

	Time UT	Latitude	Longitude
Eclipse begins	08 h 26 m 12 s	30° 19.8′ N	44° 29.1′ W
Totality begins	09 h 30 m 16 s	41° 01.5′ N	65° 04.7′ W
Mid-eclipse	11 h 03 m 16 s	45° 04.0′ N	24° 18.0′ E
Totality ends	12 h 36 m 00 s	17° 33.0′ N	87° 17.1′ E
Eclipse ends	13 h 40 m 06 s	06° 38.0′ N	68° 06.3′ E

Maximum: 1999 August 11, 11 h 3 m 15.99 s
Julian Day: 2451401.9606017
Geocentric conjunction in Right Ascension, 1999 August 11: 10 h 51 m 10.40 s
Julian Day: 2451401.9522037

Right Ascension of Sun and Moon: 9 h 23 m 6.42 s
Sun's declination: +15° 19′ 48.53″
Moon's declination: +15° 50′ 10.13″

Hourly motion:

- of the Sun in Right Ascension: 9.476s
- of the Moon in Right Ascension: 142.015s
- of thc Sun in declination: – 0′ 44.36″
- of the Moon in declination: – 7′ 42.63″

Equatorial horizontal parallax:

- of the Sun: 0′ 8.68″
- of the Moon: 58′ 44.50″

Apparent diameter of Sun: 31′ 36.60″
Apparent diameter of Moon: 32′ 0.52″

† More details of this eclipse (shadow track, times, etc.), may be found on Internet sites such as those of the British Astronomical Association (http://www.ast.cam.ac.uk/˜baa), the Bureau des Longitudes (prepared by P. Rocher, http://www.bdl.fr) and Fred Espenak (http://www.sunearth.gsfc.nasa.gov). The *Royal Greenwich Observatory Guide to the 1999 Total Eclipse of the Sun*, by Steve Bell, is available from bookshops. Observations of the solar corona are coordinated on the European level by Belgian scientist Dr. Frédéric Clette of JOSO (Joint Organization for Solar Observations), at the Royal Belgian Observatory, Avenue Circulaire 3, B-1180 Brussels.

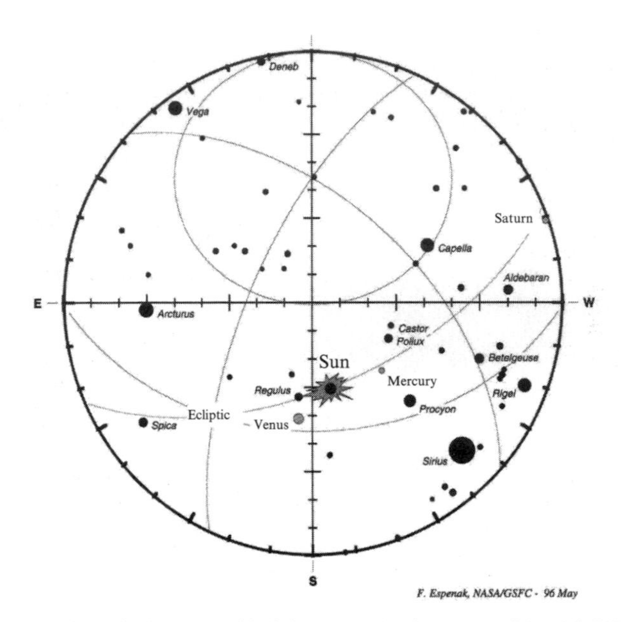

Figure D.1. Visibility of planets and bright stars, 11 August 1999, 11 h UT. (F. Espenak, NASA/GSFC.)

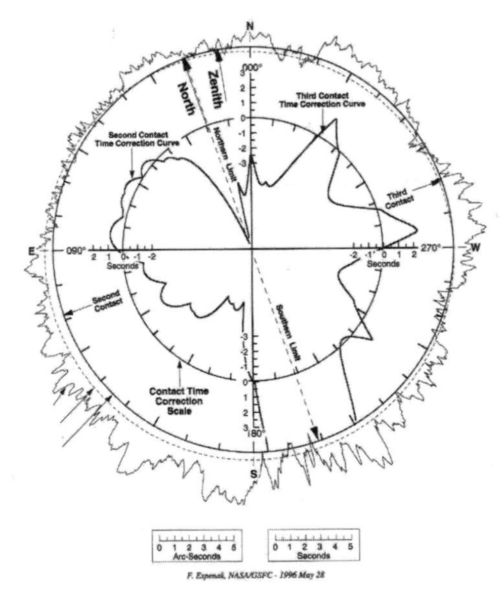

Fig. D.2. This diagram, from the NASA bulletin on the eclipse of 11 August 1999, shows the Moon's limb greatly exaggerated, the mean limb and the mean limb of the centre of mass. As well as these three limbs, a graduated hour circle, from 0° to 360°, is shown, for the purposes of orientation. Within the circle, the two sinuous curves are the time correction curves for second and third contacts. This diagram will be useful when preparing for the eclipse, and for determining, for example, the location at which Baily's Beads will be seen. (F. Espenak, NASA/GSFC.)

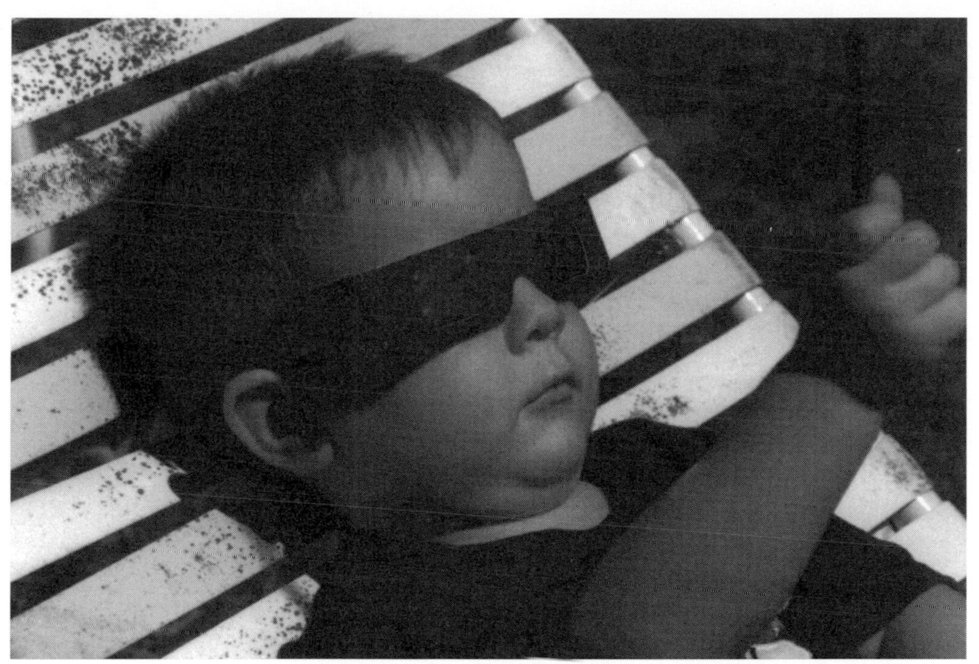

Fig. D.3. This young observer is ready to observe the partial phases of the eclipse of the Sun. It is, of course, vital to protect the sight with an adequate filter, to cut down brightness and filter out ultraviolet and infrared radiation. (V. Guillermier.)

COUNTRIES THROUGH WHICH THE TRACK PASSES

Times are in Universal Time (UT). In the following tables, the last column indicates the duration of totality if the eclipse is total for the place named, or the extent of the eclipse if partial.

Canada

(The sun is still below the horizon at the beginning of the eclipse.)

Place	Latitude (N)	Longitude (W)	Begins	Middle	Ends	Duration/ extent
Halifax	44° 38′	63° 35′	*	8 h 33 m 42 s	10 h 31 m 12 s	92%
Ile Sable	44° 00′	60° 00′	*	9 h 32 m 18 s	10 h 31 m 06 s	97%
Ottawa	45° 25′	75° 43′	*	9 h 37 m 17 s	10 h 31 m 52 s	82.3%

Saint-Pierre and Miquelon

Place	Latitude (N)	Longitude (W)	Begins	Middle	Ends	Duration/ extent
St-Pierre	46° 47′	56° 12′	8 h 39 m 30 s	9 h 35 m 06 s	10 h 35 m 18 s	91%

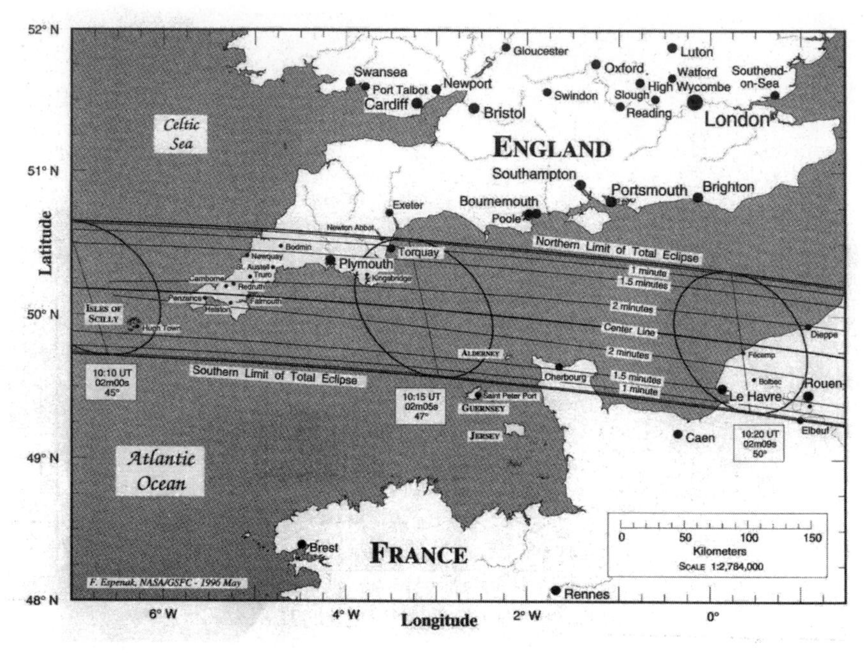

Fig. D.4. The path of the eclipse over England.

United Kingdom

Place	Latitude (N)	Longitude (W)	Begins	Middle	Ends	Duration/ extent
Falmouth	50° 09′	05° 05′	08 h 57 m 06 s	10 h 12 m 12 s	11 h 32 m 18 s	2:02
Penzance	50° 07′	05° 32′	08 h 56 m 36 s	10 h 11 m 36 s	11 h 31 m 36 s	2:02
Scilly Isles	49° 55′	06° 15′	08 h 55 m 42 s	10 h 10 m 30 s	11 h 30 m 24 s	1:42
Newquay	50° 24′	05° 06′	08 h 57 m 18 s	10 h 12 m 24 s	11 h 32 m 18 s	1:39
Plymouth	50° 23′	04° 09′	08 h 58 m 12 s	10 h 13 m 42 s	11 h 33 m 48 s	1:38
Alderney	49° 42′	02° 12′	08 h 59 m 36 s	10 h 16 m 06 s	11 h 37 m 06 s	1:38
Torquay	50° 27′	03° 31′	08:58 m 54 s	10:14 m 36 s	11:34 m 42 s	0:27
London	51° 30′	00° 05′	09:03 m 36 s	10:19 m 54 s	11:40 m 00 s	96.5%

Jersey and Guernsey

Place	Latitude (N)	Longitude (W)	Begins	Middle	Ends	Duration/ extent
Guernsey	49° 30′	02° 35′	08 h 59 m 00 s	10 h 15 m 30 s	11 h 36 m 36 s	99.8%
Jersey	49° 13′	02° 10′	08 h 59 m 12 s	10 h 16 m 00 s	11 h 37 m 24 s	99.1%

France

Place	Latitude (N)	Longitude (W/E)	Begins	Middle	Ends	Duration/ extent
Metz	49° 08′	06° 10′	09 h 09 m 12 s	10 h 29 m 00 s	11 h 51 m 30 s	2:13
Dieppe	49° 54′	01° 04′	09 h 03 m 24 s	10 h 21 m 00 s	11 h 42 m 30 s	2:04
Reims	49° 15′	04° 02′	09 h 06 m 24 s	10 h 25 m 30 s	11 h 47 m 42 s	1:59
Cherbourg	49° 40′	01° 35′	09 h 00 m 18 s	10 h 17 m 06 s	11 h 38 m 12 s	1:40
Rouen	49° 27′	01° 04′	09 h 03 m 00 s	10 h 20 m 54 s	11 h 42 m 42 s	1:40

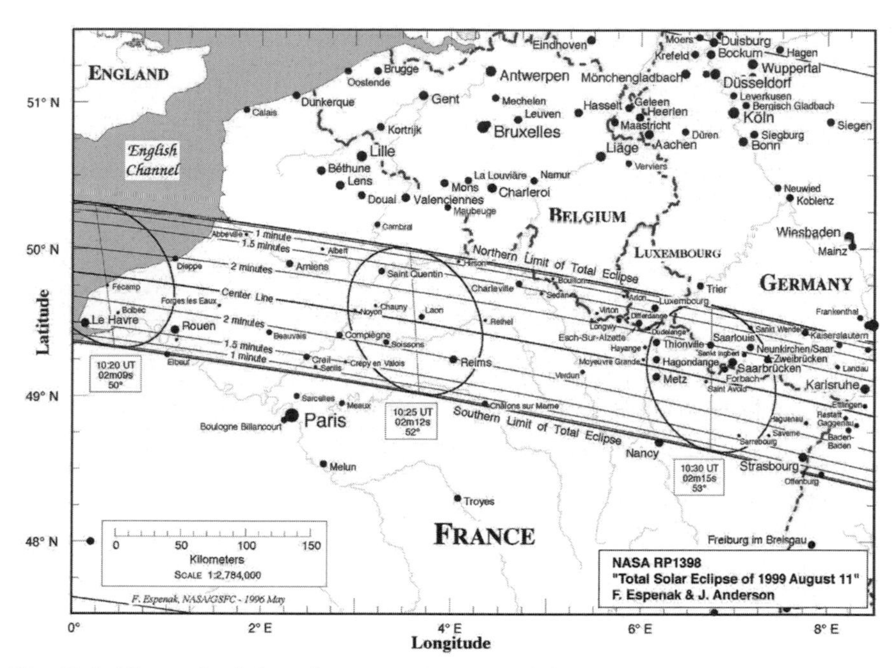

Fig. D.5. The path of the eclipse over France, Belgium, Luxembourg and Germany.

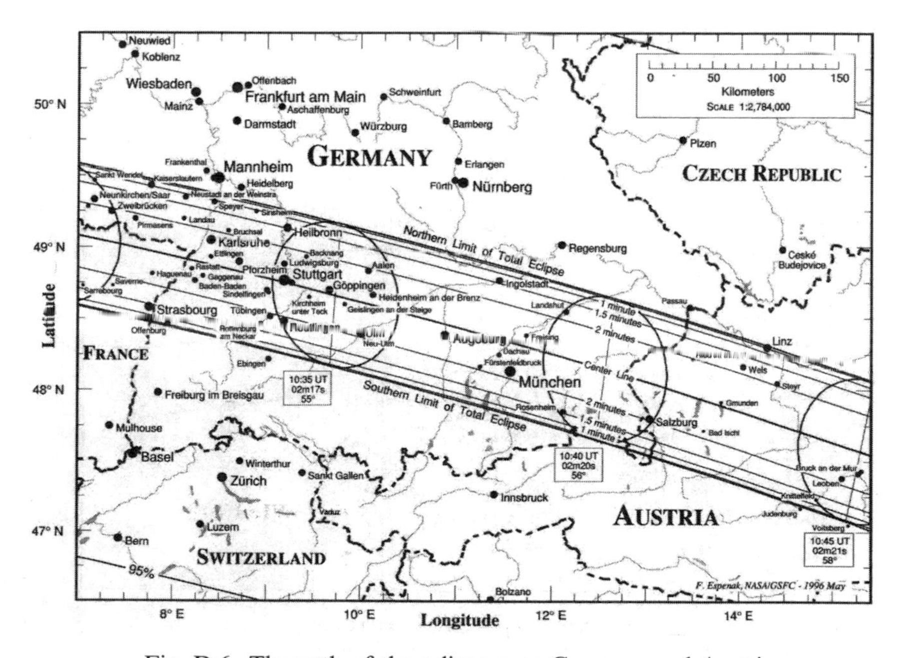

Fig. D.6. The path of the eclipse over Germany and Austria.

Place	Latitude (N)	Longitude (E)	Begins	Middle	Ends	Duration/ extent
Le Havre	49° 30′	00° 06′	09 h 02 m 00 s	10 h 19 m 30 s	11 h 41 m 06 s	1:31
Strasbourg	48° 35′	07° 45′	09 h 10 m 54 s	10 h 31 m 30 s	11 h 54 m 24 s	1:22
Nancy	48° 42′	06° 12′	09 h 09 m 00 s	10 h 29 m 06 s	11 h 51 m 54 s	0:14
Paris	48° 52′	02° 20′	09 h 04 m 06 s	10 h 22 m 42 s	11 h 45 m 06 s	99.4%

Luxembourg

Place	Latitude (N)	Longitude (E)	Begins	Middle	Ends	Duration/ extent
Luxembourg	49° 32′	05° 53′	09 h 09 m 30 s	10 h 28 m 54 s	11 h 51 m 06 s	1:14

Germany

Place	Latitude (N)	Longitude (E)	Begins	Middle	Ends	Duration/ extent
Stuttgart	48° 46′	09° 11′	09 h 13 m 06 s	10 h 33 m 54 s	11 h 56 m 42 s	2:17
Munich	48° 09′	11° 35′	09 h 16 m 18 s	10 h 38 m 12 s	12 h 01 m 18 s	2:07
Berlin	52° 32′	13° 25′	09 h 21 m 13 s	10 h 39 m 54 s	11 h 59 m 16 s	88.8%

Austria

Place	Latitude (N)	Longitude (E)	Begins	Middle	Ends	Duration/ extent
Salzburg	47° 48′	13° 03′	09 h 18 m 24 s	10 h 40 m 48 s	12 h 04 m 06 s	2:02
Steyr	48° 04′	14° 25′	09 h 20 m 42 s	10 h 43 m 12 s	12 h 06 m 06 s	1:54
Graz	47° 05′	15° 22′	09 h 22 m 00 s	10 h 45 m 24 s	12 h 08 m 42 s	1:04
Linz	48° 19′	14° 18′	09 h 20 m 30 s	10 h 42 m 48 s	12 h 05 m 30 s	0:30
Vienna	37° 29′	16° 22′	09 h 27 m 00 s	10 h 54 m 30 s	12 h 20 m 36 s	68%

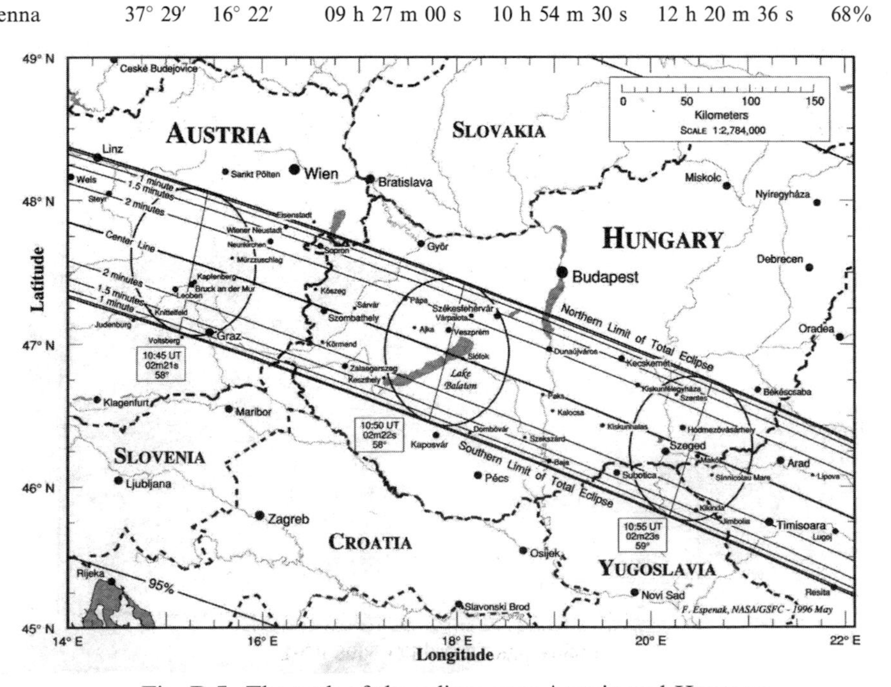

Fig. D.7. The path of the eclipse over Austria and Hungary.

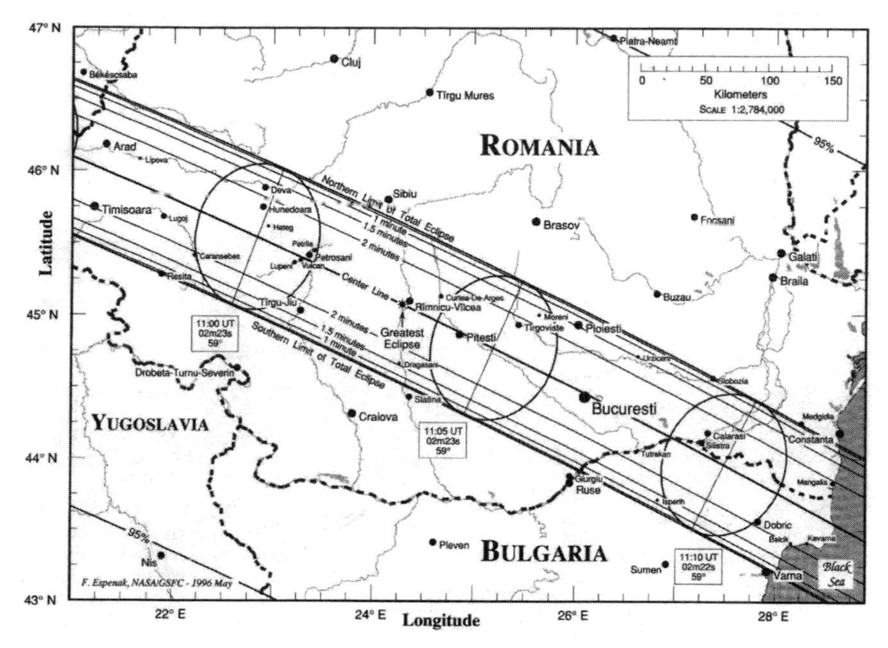

Fig. D.8. The path of the eclipse over Romania and Bulgaria.

Hungary

Place	Latitude (N)	Longitude (E)	Begins	Middle	Ends	Duration/ extent
Szombathely	47° 14′	16° 38′	09 h 24 m 06 s	10 h 47 m 36 s	12 h 10 m 36 s	2:22
Budapest	47° 29′	19° 05′	09 h 28 m 12 s	10 h 51 m 42 s	12 h 14 m 00 s	99.4%

Romania

Place	Latitude (N)	Longitude (E)	Begins	Middle	Ends	Duration/ extent
Bucharest	44° 26′	26° 05′	09 h 41 m 24 s	11 h 06 m 54 s	12 h 28 m 42 s	2:22
Silistra	44° 07′	27° 17′	09 h 43 m 54 s	11 h 09 m 24 s	12 h 30 m 48 s	2:21

The maximum duration of the eclipse (2 minutes 23 seconds) occurs at Latitude 45° 04′ N, Longitude 24° E.

Bulgaria

Place	Latitude (N)	Longitude (E)	Begins	Middle	Ends	Duration/ extent
Shabla	43° 31′	28° 32′	09 h 46 m 36 s	11 h 12 m 24 s	12 h 33 m 30 s	2:21
Sofia	42° 41′	23° 19′	09 h 37 m 00 s	11 h 03 m 48 s	12 h 27 m 18 s	94.4%

Appendix E

Eclipses of the Sun and Moon until 2010

The following tables give the date, type of eclipse, times, duration and zones of visibility. (For further details, see *Eclipse!* by P.S. Harrington, which describes, with numerous maps and tables, every eclipse of the Sun and Moon until 2017.)

TOTAL, ANNULAR AND PARTIAL ECLIPSES OF THE SUN

Times are given in UT.

Date	Type	Zone of visibility	Time of maximum	Duration of totality/ Maximum phase
26 Feb 1998	Total	Galapagos, Colombia Venezuela, Caribbean	17 h 29	04 m 09 s
22 Aug 1998	Annular	Sumatra, Borneo	02 h 07	03 m 14 s
16 Feb 1999	Annular	Australia	06 h 35	01 m 19 s
11 Aug 1999	Total	Europe, S. Asia, India	11 h 04	02 m 23 s
5 Feb 2000	Partial			58%
1 Jul 2000	Partial			48%
31 Jul 2000	Partial			60%
25 Dec 2000	Partial			72%
21 Jun 2001	Total	Angola, Mozambique Madagascar	12 h 04	04 m 56 s
14 Dec 2001	Annular	Pacific, C. America	20 h 54	03 m 53 s
10 Jun 2002	Annular	Pacific	23 h 45	00 m 23 s
4 Dec 2002	Total	Angola, Mozambique Australia	07 h 33	02 m 04 s
31 May 2003	Annular	Greenland	04 h 10	03 m 37 s
23 Nov 2003	Total	Antarctic	22 h 47	01 m 57 s
19 Apr 2004	Partial			74%
14 Oct 2004	Partial			93%
8 Apr 2005	Ann/Tot	S. Pacific, C. America	20 h 42	00 m 42 s

Date	Type	Zone of visibility	Time of maximum	Duration of totality/ Maximum phase
3 Oct 2005	Annular	Spain, Algeria, Libya, Ethiopia	10 h 28	00 m 42
29 Mar 2006	Total	Africa, Middle East, Russia	10 h 15	04 m 07 s
22 Sep 2006	Annular	Brazil, S. Atlantic	11 h 40	07 m 09 s
19 Mar 2007	Partial			88%
11 Sep 2007	Partial			75%
7 Feb 2008	Annular	Antarctic	03 h 55	02 m 12 s
1 Aug 2008	Total	Alaska, Greenland, Russia, China	10 h 25	02 m 27 s
26 Jan 2009	Annular	Borneo, Sumatra, Indian Ocean	07 h 57	07 m 54 s
22 Jul 2009	Total	India, Nepal, China	02 h 36	06 m 39 s
15 Jan 2010	Annular	C. Africa, India, Burma, China	07 h 10	11 m 07 s
11 Jul 2010	Total	S. Pacific	19 h 30	05 m 20 s

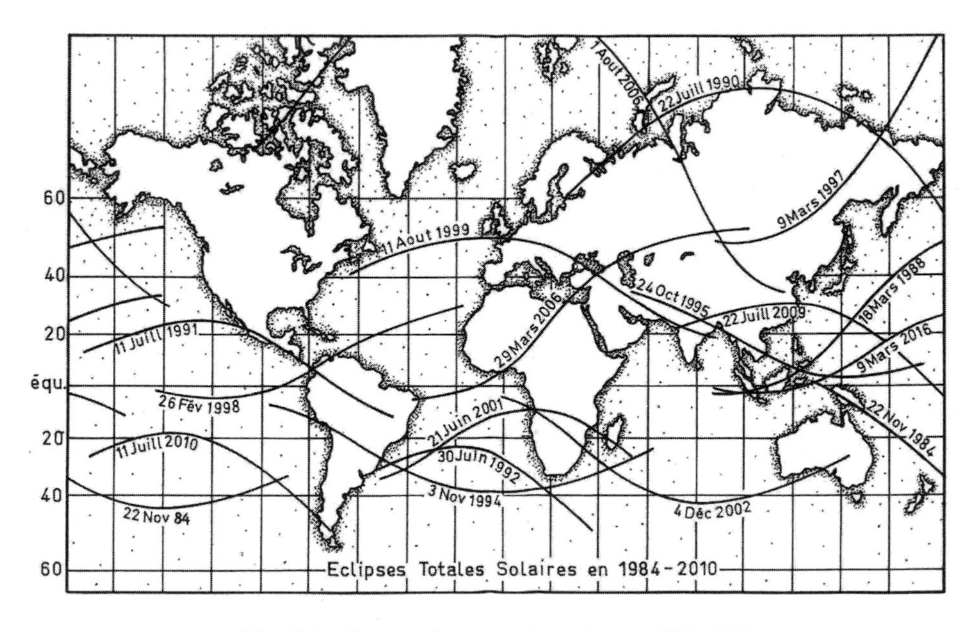

Fig. E.1. Tracks of total solar eclipses, 1984–2010.

ECLIPSES OF THE MOON

Times are given in UT.

Date	Magnitude (%)	Partial phase begins	Totality begins	Mid-eclipse	Totality ends	Partial phase ends	Zone of visibility
28 Jul 1999	40	10 h 22	–	11 h 33	–	12 h 44	Pacific only
21 Jan 2000	133	03 h 05	04 h 08	04 h 46	05 h 24	06 h 27	France and America
16 Jul 2000	177	11 h 58	13 h 03	13 h 56	14 h 49	15 h 54	Australia, Pacific
9 Jan 2001	118	18 h 43	19 h 50	20 h 20	20 h 50	21 h 57	Europe, Africa, Asia
5 Jul 2001	49	13 h 39	–	14 h 58	–	16 h 17	Partial Australia only
16 May 2003	113	02 h 06	03 h 17	03 h 43	04 h 09	05 h 20	Beginning visible in France, total S. America
09 Nov 2003	102	23 h 35	01 h 09	01 h 20	01 h 31	03 h 05	Europe
04 May 2004	130	18 h 49	19 h 53	20 h 30	21 h 07	22 h 11	End visible in France, total C. Europe
28 Oct 2004	131	01 h 15	02 h 24	03 h 04	03 h 44	04 h 53	France and America
17 Oct 2005	6	11 h 34	–	12 h 02	–	12 h 30	Pacific only
07 Sep 2006	18	18 h 10	–	18 h 54	–	19 h 38	End visible in France, seen from C. Europe
03 Mar 2007	123	21 h 33	22 h 47	23 h 23	23 h 59	01 h 13	France
28 Aug 2007	147	08 h 51	09 h 51	10 h 36	11 h 21	12 h 21	Pacific only
21 Feb 2008	111	01 h 46	03 h 03	03 h 28	03 h 53	05 h 10	Europe, America
16 Aug 2008	81	19 h 36	–	21 h 09	–	22 h 42	Beginning visible in France, total in Africa
31 Dec 2009	7	18 h 57	–	19 h 25	–	19 h 53	Europe, Africa, Asia
26 Jun 2010	53	10 h 18	–	11 h 39	–	13 h 00	Pacific only
21 Dec 2010	125	06 h 34	07 h 41	08 h 17	08 h 53	10 h 00	Beginning visible France, total N. America

Addresses and bibliography

MAGAZINES AND JOURNALS

The covers of some magazines, which can be a mine of information on current events in astronomy.

Aster Spanish language, Agrupacio Astronomica de Barcelona, Passeig de Gracia, 71 atic, E-08008 Barcelona, Spain.

Astronomie-Québec In French. 4545 Ave. Pierre de Coubertin, CP 1000 Succursal M, Montréal, Québec, H1V3R2 Canada.

Astronomy American monthly. 21027 Crossroads Circle, PO Box 1612, Waukesha, WI 53187, USA. http://www.kalmbach.com/astro/astronomy.html

Astronomy Now UK monthly. AN Subscription Dept., Freepost, London W12 9BR, UK.

Journal of the British Astronomical Association Subscription and membership enquiries to: The Assistant Secretary, British Astronomical Association, Burlington House, Piccadilly, London W1 V 9AG.

Ciel et Espace French monthly. Association Française d'Astronomie. 17, Rue Emile Deutsch-de-la-Meurthe, 75014 Paris, France.

Ciel et Terre Royal Belgian Astronomical Society, Avenue Circulaire, B-1180 Brussels, Belgium.

Eclipse Société Objectif Véga. In French. 93147 Bondy Cedex, France.

Il Cielo Italian monthly. Biroma Editore, Via San Pio X, 108, 35015 Galliera Veneta, Italy.

L'Astronomia Italian monthly. Edizione Media Presse s.r.l., Via Nono Bixio, 30, 20129 Milano, Italy.

L'Astronomie Monthly magazine of the Société Astronomique de France (SAF), founded by Camille Flammarion in 1882. SAF, 3 Rue Beethoven, 75016 Paris, France.

Le Ciel Liège Astronomical Society, Institut d'Astrophysique, Avenue de Cointe 5, 4000 Liège, Belgium.

Orion Swiss Astronomical Society. In French. SAS, Paul-Emile Muller, Ch. Marais-Long 10, 1217 Meyrin, Switzerland.

Pulsar Société d'Astronomie Populaire. SAP, 1 Avenue Camille-Flammarion, 31500 Toulouse, France.

Sky and Telescope American monthly. Sky and Telescope, PO Box 9111, Belmont, MA 02138-9111, USA. http://www.skypub.com

Southern Astronomy Australian. Subscriptions Dept., 116 Bronte Road, Bondi Junction 2022, Sydney, NSW, Australia.

TRAVEL AGENCIES OFFERING ECLIPSE TRIPS

Explorers Tours, 223 Coppermill Road, Wraysbury TW19 5NW Tel: 01753 681999 Fax: 01753 682660

Scientific Expeditions Inc., 227 West Miami Avenue, Suite 3, Venice, FL 34285, USA

SITA Travel, 16250 Venura Blvd. #300, Encino, CA 91436-2211, USA

Outer Edge Expeditions, 45,500 Pontiac Tr, Walled Lk, MI 48390, USA

For further information, see the advertisement sections of *Astronomy Now, Sky and*

Telescope, and *Astronomy*, which feature many advertisements for astronomical travel. In France, the AFA and SAF sometimes organise 'theme' trips; for example, to the Baikonur Cosmodrome or to observe an eclipse of the Sun.

SOME INTERNET ADDRESSES

AstroWeb	http://fits.cv.nrao.edu/www/astronomy.html http://marvel.stsci.edu/net-resources.html
NASA Goddard Space Flight Center	http://www.gsfc.nasa.gov
NASA Jet Propulsion Laboratory	http://www.jpl.nasa.gov/
NASA Solar Information, site SOHO	http://sohowww.nascom.nasa.gov/
NASA Solar Eclipse Bulletins	http://sunearth.gsfc.nasa.gov/sdac.html
European Southern Observatory	http://www.hq.eso.org/
Skylink	http://www2.ec-lille.fr/~astro
Centre National d'Études Spatiales	http://www.cnes.fr
Bureau des Longitudes	http://www.bdl.fr
Institut d'Astrophysique de Paris	http://www.iap.fr
Institut d'Astrophysique Spatiale	http://www.ias.fr
Institut de Radioastronomie Millimétrique	http://www.iram.fr
Observatoire de Haute-Provence	http://www.obs-hp.fr
Observatoire de la Côte d'Azur	http://www.obs-azur.fr
Observatoire de Montréal	http://www.astro.umontreal.ca/home/
Observatoire de Nice	http://www.obs-nice.fr
Observatoire de Paris	http://www.obspm.fr
Observatoire Midi-Pyrénées	http://www.obs-mip.fr
Observatoire de Genève	http://www.unige.ch
Big Bear Solar Observatory	http://bigbear.caltech.edu/cgi-bin/daily.cgi
Kitt Peak Observatory	http://argo.tuc.noao.edu:2001/synoptic.html
Sacramento Peak Observatory	http://www.sunspot.noao.edu/index.html
Wilcox Solar Observatory	http://quake.stanford.edu/~wso/wso.html
Yohkoh	http://porel.space.lockheed.com:80/STX/html2 http://isasxa.solar.isas.ac.jp/~freeland
IAU Working Group on Eclipses	http://www.williams.edu/Astronomy/IAU_eclipses
Espenak at NASA	http://sunearth.gsfc.nasa.gov
SOHO Synoptic Database	http://orpheus.nascom.nasa.gov/synoptic
Satellite SOHO	http://umtof-umd.edu.pm/
Société Astronomique de France	http://www.iap.fr/saf
British Astronomical Association	http://www.ast.cam.ac.uk/~baa

NASA (NATIONAL AERONAUTICS AND SPACE ADMINISTRATION)

NASA, Goddard Space Flight Center, Code 693, Greenbelt, MD 20771, USA. From this address can be obtained NASA publications about past or future eclipses by sending an A4 self-addressed envelope, with appropriate postage in international postal coupons (310 g) for the attention of Fred Espenak. The envelope should be marked in the top left corner with the month and year of the eclipse concerned. These documents are also available on the Internet site of the Solar Data Analysis Center (http://sunearth.gsfc.nasa.gov/sdac.html, and http://sunearth.gsfc.nasa.gov/eclipse/predictive/eclipse-path.html).

PROGRAMS FOR SIMULATION AND IMAGE PROCESSING

Below is a list of some of the software available for simulation and image processing.

Simulation software

In English:

Dance of the Planets ARC Science Simulations, PO Box 1955A, Loveland CO 80539 USA. For PCs. Solar System simulation.

The Sky Software Bisque, 912 Twelfth Street, Suite A, Golden CO 80401, USA. For PCs.

RedShift 3 Maris Multimedia Ltd., 99 Mansell Street, London E1 8AX. e-mail: Maris@maris.com PC or Macintosh.

Distant Suns Virtual Reality. PC or Macintosh.

In French:

Eclipwin (eclipse calculation and simulation)

Ephémérides (complete ephemerides for planets, comets and asteroids)

Cadsol (sundials)

Albiréo (sky chart for beginners)

Planétaire (ephemerides of Moon and planets, events for given dates. Sundials)

All the above French versions are available from l'Association française d'astronomie, 17 Rue Emile Deutsch-de-la-Meurthe, 75014 Paris, France.

Astronomie 3000 (Sky Plot Prò. Micro Application Soft. Calculations, astronomy courses, NASA images)

Image processing software

Editim Plus CCD image processing/Windows. Suitable for most commercial CCD cameras (Formats: ST4, ST6, HI-SIS, FITS, BMP, PCX, IMG, IMA).

PhotoShop Adobe Systems. PC or Macintosh. Powerful image processing program.

PhotoStyler Aldus Corp. PC. General image processing program.

Imagine-32 CompuScope, 3463 State St., Suite 431, Santa Barbara, CA 93105, USA. PC. 32-bit image processing.

SkyPro Software Bisque, 912 Twelfth St., Suite A, Golden, CO 80401, USA. PC. Astronomical image processing.

PaintShop Pro JASC Inc., 10901 Red Circle Drive, Suite 340, Minnetonka, MN 55343, USA. General image processing program.

Hidden Image Seghal Corp., 4 Sayre Drive, Princeton, NJ 08540, USA. Powerful astro-image processing program; maximum entropy deconvolution, Fourier filter, image calibration, unsharp mask,; formats: SBIG, FITS, TIFF, Lynxx, etc. PC.

QMiPS Christian Buil and his team have devised this program, offering all analysis functions, processing for astronomical images, all CCD image formats.

Mira Axiom Research, 2450 E. Speedway, Ste. 3, Tucson, AZ 85719, USA. PC. Powerful CCD image treatment program.

PLANETARIA IN THE BRITISH ISLES

This list, kindly provided by the British Association of Planetaria, gives details of planetaria in the British Isles. There are also many travelling domes, details of which may be obtained from the addresses at the end. This is not an exhaustive list.

Aberdeen College of Further Education, Gallowgate, Aberdeen AB9 1DN (01224 640366)

Armagh Planetarium, College Hill, Armagh, N. Ireland BT61 9DB (01861 523689)

Bristol Exploring Starlab, Templemeads, Bristol BS1

Chigwell School, High Road, Chigwell, Essex IG7 6QF (0181 500 4106)

Dublin Science Expo, Dunsink Observatory, Dublin 15, Ireland (010 3531 668 0748)

Dundee: Mills Observatory, Balgay Park, Dundee DD2 2UB (01382 67138)

Edinburgh Royal Observatory, Blackford Hill, Edinburgh EH9 3HJ (0131 668 8406)

Glasgow College of Nautical Studies, 21 Thistle Street, Glasgow G5 9XR

Glasgow University Observatory, Acre Road, Glasgow G20 0TL 0141 946 5213

Helston Planetarium Project, Orchard Lawn Farm, Lr. Porkellis, TR13 0JT (01326 341117)

Jodrell Bank Science Centre, Lr. Withington, Macclesfield SK11 9DL (01477 571339)

Liverpool Museum, William Brown Street, Liverpool; L3 8EN (0151 207 0001)

London Planetarium, Marylebone Road, London NW1 5LR (0171 935 6861)

London Caird Planetarium, Old Royal Observatory, Greenwich SE10 9NF (0181 858 4422)

Plymouth: William Day Planetarium, University of Plymouth, Drake Circus, Plymouth PL4 8AA (01752 232462)

Powys Planetarium, Lanshay Lane, Knighton, Powys LD7 1LW (01547 520247)

Royal Masonic School Planetarium, Chorley Wood Road, Rickmansworth, Herts WD5 4HF

Schull Planetarium, Schull Community College, Colla Road, Schull, Co. Cork, Ireland (00353 21 028 28552)

Sheffield Astrosphere, 188 Gleadless Common, Sheffield S12 2US (0114 265 9585)
Sidmouth: Norman Lockyer Observatory, Windwood, Higher Brook Meadow, Sidford EX10 9SS
Southend Planetarium, Central Museum, Victoria Avenue, Southend SS2 6EW (01702 434449)
Southampton University, Faculty of Physics, Southampton SO9 5NH
South Tyneside College, St George's Avenue, South Shields, Tyne and Wear NE34 6ET (0191 427 3589)
Todmorden Planet Earth Centre, Bacup Road, Todmorden, Lancs OL14 7HW (01706 816964)
Yarmouth (IOW): Fort Victoria Planetarium, Westhill Lane, Norton, Yarmouth, IOW PO41 0RR (01983 761555)

Travelling planetaria
Aylesford Skylab: Murray Barber, 14 Cedar Close, Ditton, Aylesford, Kent ME20 6EN (01622 719725)
Mizar Travelling Planetarium: Bob Mizon, 38 The Vineries, Colehill, Wimborne, Dorset BH21 2PX (01202 887084)
StarDome (NIAS) Planetarium: Frank Gear, 251 Abington Avenue, Northampton NN3 2BU (01604 714712)

Bibliography

GENERAL ASTRONOMY

À l'affût des étoiles, P. Bourge, et J. Lacroux, Dunod 14e édition, 1997.
Astronomie générale – Astronomie sphérique et éléments de mécanique céleste, A. Danjon, Albert Blanchard, 2e édition, 1980.
Astronomie – Le guide d l'observateur, P. Martinez (sous la direction de), 2 vol., Édition Association Adagio 10, rue A. Daudet, 31200 Toulouse, 1986.
Astronomie, J.-C. Pecker (sous la direction de), 2 vol., Flammarion, 1985.
Astronomie, Ph. de La Cotardière (sous la direction de), Larousse, 1994.
Astrophysique générale, J.-C. Pecker et E. Schatzman, Masson, 1959.
Ciel et astronomie, passion, M. Marcelin, Hachette, 1986.
Encyclopédie d'astronomie de Cambridge, le Grand Atlas Universalis de l'astronomie J. Audouze et G. Israël (sous la direction de), Encyclopedia Universalis, 1983.
Initiation à l'astronomie, A. Acker, Masson 4e édition, 1990.
Introduction à l'astronomie, A. Acker, Masson, 1997.
Les Guides du ciel 1995–1996 et 1996–1997, G. Cannat, Nathan.
Physik der Sternatmosphären, A. Unsöld, Springer, Berlin and Heidelberg, 1955.
Solar Interior and Atmosphere, A.N. Cox. W.C. Livingston and M.S. Matthews, *Ed. The University of Arizona Press, Tucson, 1991.*
Voyage dans le Système solaire, S. Brunier, Bordas, 1993.

THE MOON

Atlas-guide photographique de la Lune, G. Viscardy, Association franco-monégasque d'astronomie, Masson, 1985.
De la Terre à la Lune, J. Verne, édition Jules Hetzel & Cie, 1870.
Le livre de la Lune, J.-L. Heudier, Z'éditions, 1996.
Les vies des nobles grecs et romains, Plutarque.
Seven Pillars of Wisdom, T.E. Lawrence, 1926.
The Fall of Constantinople, S. Ruciman, Cambridge University Press, 1965.

THE SUN

Astrophysics of the Sun, H. Zirin, Cambridge University Press, 1988.
Guide to the Sun, K. Phillips, New York, Cambridge University Press, 1992.
Illustrated Glossary for Solar and Solar–Terrestrial Physics, edited by Bruzek and Durrant, Reidel Pub. Comp., 1977.
L'avenir du Soleil, J.-C. Pecker, Hachette, 1990.
La Lumière des neutrinos, M. Cribier, M. Spiro et D. Vignaud, Seuil, 1995.
Le Soleil est une étoile, J.-C. Pecker, Presses Pocket, 1992.
Le Soleil, G. Bruhat, Librairie F. Alcan, 1931.
Le Soleil, P. Lantos, Collection Que sais-je? PUF, 1994.
Le Soleil, Le P.A.S.-J. Secchi, 2 vol., (2e édition), Gauthier-Villars, Paris, 1875.
Le Soleil, R. Michard, Collection Que sais-je? No 230, PUF, 1966.
Nostalgie de la lumière, M. Cassé, Belfond, 1987.
Observing the Sun, P. Taylor, New York, Cambridge University Press, 1991.
Our Sun, D.H. Menzel, Harvard University Press, 1959.
Secrets of the Sun, R. Giovanelli, New York, Cambridge University Press, 1984.
Solar and Stellar Activity Cycles, P.R. Wibson, Cambridge University Press, 1994.
Solar Astr. Handbook, R. Beck, H. Hilbrecht, K. Reinsch and P. Völker, Willmann–Bell Inc, 1995.
Solar Astrophysics, P. Foukal, John Wiley and Sons, NY, 1990.
Solar Interior and Atmosphere, A. Cox, W. Livingston and M. Mathews, University of Arizona Press, 1991.
Solar magnetohydrodynamics, E. Priest, Riedel, 1982.
Sous l'étoile Soleil, J.-C. Pecker, Fayard, 1984.
Sun, Earth and Sky, K. Lang, Springer, 1997.
The Restless Sun, D. Wentzel, Smithsonian Institution Press, Washington DC, 1982.
The Sun, M. Stix, Springer-Verlag, Berlin, 1989.
The Sun, our star, R.W. Noyes, Harvard University Press, 1982.

ECLIPSES

Annuaire du Bureau des Longitudes, C. Arago, 1842.
Astronomical Formulae For Amateur, J. Meeus, William Bell, 1982.
Calculs astronomiques, J. Meeus, Société astronomique de France, 1986.
Canon of Eclipses, R.T. Von Oppolzer, New York, Dover Publications, 1962.
Canon of Solar Eclipses – 2003 to + 2526 et Canon of Lunar Eclipses – 2002 to + 2526, H. Mucke et J. Meeus, Astronomische Büro, Vienne, 2e édition, 1983.
Chasing the shadow: an observer's guide to solar eclipses, J. Harris and R. Talcott, Kalmbach Books, 1994.
Communication à l'Académie des Sciences du 10 mai 1683, Cassini.
Connaissances des Temps, Bureau des longitudes, Gauthier-Villars.
Eclipse! P.S. Harrington, John Wiley and Sons Inc., 1997.
Eclipses of the Sun, S.A. Mitchell, New York, Columbia University Press, 1923.

Einstein. La joie de la pensée, F. Balibar, Gallimard, Découvertes-sciences, 1993.
Éléments d'astronomie fondamentale, M. Danloux-Dumesnil, Albert Blanchard, 1985.
Éphémérides Astronomiques, Annuaire du Bureau des Longitudes, Gauthier-Villars, annuel depuis 1795.
Fifty Year Canon of Solar Eclipses: 1986–2035, F. Espenak, Cambridge, MA. Sky Publishing Corporation, Sky-NASA Reference Publication 1178 – July 1987.
La Bible, l'Évangile selon Saint Jean, commentaires des professeurs de la Faculté de théologie de l'Université de Navarre (Les Éditions du Laurier, 1996.
Les aventures de Tintin. Le temple du Soleil, Hergé, Casterman, 1949.
Les éclipses, P. Couderc, PUF, collection Que sais-je? épuisé.
Observe Eclipses, M.D. Reynolds, R.A. Sweetsir, The Astronomical League, 2e édition, 1995.
Œuvres scientifique de J. Janssen, Tomes I et II, Paris, 1929.
The Cambridge Eclipse Photography Guide, J.M. Pasachoff, M.A. Covington, Cambridge University Press, 1993.
The Sun in Eclipse, M. Maunder and P. Moore, Cambridge University Press, 1997.
Total Eclipses of the Sun, J.B. Zirker, Princeton University Press, 1995.
Traité physique et historique de l'aurore boréale, De Mairan, Paris, Imprimerie royale, 1733.
UK Solar Eclipses From Year 1, S. Williams, Clock Tower Press, 1996.

THE SOLAR CORONA

A Guide to the Solar Corona, D.F. Billings, Ac. Press Ed., 1966.
Interplanetary dynamical processes, E.N. Parker, Interscience Publication, 1963.
Introduction to the Solar Wind, J.C. Brandt, Ed. Freeman and Comp., 1970.
La couronne solaire et l'émission corpusculaire dans l'espace intreplanétaire, S.K. Vsekhsvjatskij et coll, en russe aux Éditions universitaires de Kiev, 1965.
Physics of the Solar Corona, I.S. Shklovsky, Pergamon Press, 1961.
Solar Coronal Structures, IAU coll. 144, Ed. V. Rusin, P. Heinzel et J.C. Vial, Veda Publication Slovak Academic Science, 1994.
The Solar Corona, L. Golub, J.M. Pasachoff, Cambridge University Press, 1997.
A New Sun, J.A. Eddy, NASA, SP-402.

ASTROPHOTOGRAPHY

Astrophotography for the amateur, M.A. Covington, Cambridge University Press, 1998.
La photographie astronomique d'amateur, P. Bourge, J. Dragesco et Y. Dargery, Paul Montel, 3e édition, 1987.
Le guide de l'astronomie CCD, P. Martinez et A. Klotz, Association Adagio 10, rue A. Daudet, 31200 Toulouse.

Index